ELEMENTS OF RANDOM WALK AND DIFFUSION PROCESSES

Wiley Series in
Operations Research and Management Science

Operations Research and Management Science (ORMS) is a broad, interdisciplinary branch of applied mathematics concerned with improving the quality of decisions and processes and is a major component of the global modern movement towards the use of advanced analytics in industry and scientific research. The *Wiley Series in Operations Research and Management Science* features a broad collection of books that meet the varied needs of researchers, practitioners, policy makers, and students who use or need to improve their use of analytics. Reflecting the wide range of current research within the ORMS community, the Series encompasses application, methodology, and theory and provides coverage of both classical and cutting edge ORMS concepts and developments. Written by recognized international experts in the field, this collection is appropriate for students as well as professionals from private and public sectors including industry, government, and nonprofit organization who are interested in ORMS at a technical level. The Series is comprised of three sections: Decision and Risk Analysis; Optimization Models; and Stochastic Models.

Advisory Editors • Stochastic Models

Raúl Gouet, University of Chile
Tava Olsen, The University of Auckland

Founding Series Editor

James J. Cochran, Louisiana Tech University

Decision and Risk Analysis

Barron • *Game Theory: An Introduction,* Second Edition

Forthcoming Titles

Nussbaum and Mislick • *Cost Estimation: Methods and Tools Optimization Models*

Optimization Models

Ghiani, Laporte, and Musmanno • *Introduction to Logistics Systems Management,* Second Edition

Stochastic Models

Ibe • *Elements of Random Walk and Diffusion Processes*

ELEMENTS OF RANDOM WALK AND DIFFUSION PROCESSES

OLIVER C. IBE

University of Massachusetts
Lowell, Massachusetts

Library of Congress Cataloging-in-Publication Data:

Ibe, Oliver C. (Oliver Chukwudi), 1947–
    Elements of random walk and diffusion processes / Oliver C. Ibe, University of
Massachusetts Lowell, MA.
        pages   cm
    Includes bibliographical references and index.
    ISBN 978-1-118-61809-7 (hardback)
1. Random walks (Mathematics)   2. Diffusion processes.   I. Title.
    QA274.73.I24 2013
    519.2′82–dc23

        2013009389

Printed in the United States of America

ISBN: 9781118618097

10 9 8 7 6 5 4 3 2 1

CONTENTS

PREFACE   **xiii**

ACKNOWLEDGMENTS   **xv**

1   REVIEW OF PROBABILITY THEORY   **1**

   1.1   Introduction   1

   1.2   Random Variables   1

       1.2.1   Distribution Functions   2

       1.2.2   Discrete Random Variables   3

       1.2.3   Continuous Random Variables   3

       1.2.4   Expectations   4

       1.2.5   Moments of Random Variables and the Variance   4

   1.3   Transform Methods   5

       1.3.1   The Characteristic Function   5

       1.3.2   Moment-Generating Property of the Characteristic Function   6

       1.3.3   The s-Transform   6

       1.3.4   Moment-Generating Property of the s-Transform   7

       1.3.5   The z-Transform   7

       1.3.6   Moment-Generating Property of the z-Transform   8

   1.4   Covariance and Correlation Coefficient   9

1.5	Sums of Independent Random Variables	10
1.6	Some Probability Distributions	11
	1.6.1 The Bernoulli Distribution	11
	1.6.2 The Binomial Distribution	12
	1.6.3 The Geometric Distribution	12
	1.6.4 The Poisson Distribution	13
	1.6.5 The Exponential Distribution	13
	1.6.6 Normal Distribution	14
1.7	Limit Theorems	16
	1.7.1 Markov Inequality	16
	1.7.2 Chebyshev Inequality	17
	1.7.3 Laws of Large Numbers	17
	1.7.4 The Central Limit Theorem	18
	Problems	19

2 OVERVIEW OF STOCHASTIC PROCESSES — **21**

2.1	Introduction	21
2.2	Classification of Stochastic Processes	22
2.3	Mean and Autocorrelation Function	22
2.4	Stationary Processes	23
	2.4.1 Strict-Sense Stationary Processes	23
	2.4.2 Wide-Sense Stationary Processes	24
2.5	Power Spectral Density	24
2.6	Counting Processes	25
2.7	Independent Increment Processes	25
2.8	Stationary Increment Process	25
2.9	Poisson Processes	26
	2.9.1 Compound Poisson Process	28
2.10	Markov Processes	29
	2.10.1 Discrete-Time Markov Chains	30
	2.10.2 State Transition Probability Matrix	31
	2.10.3 The k-Step State Transition Probability	31
	2.10.4 State Transition Diagrams	32
	2.10.5 Classification of States	33
	2.10.6 Limiting-State Probabilities	34
	2.10.7 Doubly Stochastic Matrix	35
	2.10.8 Continuous-Time Markov Chains	35
	2.10.9 Birth and Death Processes	36

2.11   Gaussian Processes   38
2.12   Martingales   38
       2.12.1   Stopping Times   40
       Problems   41

3   ONE-DIMENSIONAL RANDOM WALK   44

3.1   Introduction   44
3.2   Occupancy Probability   46
3.3   Random Walk as a Markov Chain   49
3.4   Symmetric Random Walk as a Martingale   49
3.5   Random Walk with Barriers   50
3.6   Mean-Square Displacement   50
3.7   Gambler's Ruin   52
       3.7.1   Ruin Probability   52
       3.7.2   Alternative Derivation of Ruin Probability   54
       3.7.3   Duration of a Game   55
3.8   Random Walk with Stay   56
3.9   First Return to the Origin   57
3.10   First Passage Times for Symmetric Random Walk   59
       3.10.1   First Passage Time via the Generating Function   59
       3.10.2   First Passage Time via the Reflection Principle   61
       3.10.3   Hitting Time and the Reflection Principle   64
3.11   The Ballot Problem and the Reflection Principle   65
       3.11.1   The Conditional Probability Method   66
3.12   Returns to the Origin and the Arc-Sine Law   67
3.13   Maximum of a Random Walk   72
3.14   Two Symmetric Random Walkers   73
3.15   Random Walk on a Graph   73
       3.15.1   Proximity Measures   75
       3.15.2   Directed Graphs   75
       3.15.3   Random Walk on an Undirected Graph   76
       3.15.4   Random Walk on a Weighted Graph   80
3.16   Random Walks and Electric Networks   80
       3.16.1   Harmonic Functions   82
       3.16.2   Effective Resistance and Escape Probability   82
3.17   Correlated Random Walk   85
3.18   Continuous-Time Random Walk   90
       3.18.1   The Master Equation   92

3.19	Reinforced Random Walk	94
	3.19.1 Polya's Urn Model	94
	3.19.2 ERRW and Polya's Urn	96
	3.19.3 ERRW Revisited	97
3.20	Miscellaneous Random Walk Models	98
	3.20.1 Geometric Random Walk	98
	3.20.2 Gaussian Random Walk	99
	3.20.3 Random Walk with Memory	99
3.21	Summary	100
	Problems	100

4   TWO-DIMENSIONAL RANDOM WALK   **103**

4.1	Introduction	103
4.2	The Pearson Random Walk	105
	4.2.1 Mean-Square Displacement	105
	4.2.2 Probability Distribution	107
4.3	The Symmetric 2D Random Walk	110
	4.3.1 Stirling's Approximation of Symmetric Walk	112
	4.3.2 Probability of Eventual Return for Symmetric Walk	113
	4.3.3 Mean-Square Displacement	114
	4.3.4 Two Independent Symmetric 2D Random Walkers	114
4.4	The Alternating Random Walk	115
	4.4.1 Stirling's Approximation of Alternating Walk	117
	4.4.2 Probability of Eventual Return for Alternating Walk	117
4.5	Self-Avoiding Random Walk	117
4.6	Nonreversing Random Walk	121
4.7	Extensions of the NRRW	126
	4.7.1 The Noncontinuing Random Walk	126
	4.7.2 The Nonreversing and Noncontinuing Random Walk	127
4.8	Summary	128

5   BROWNIAN MOTION   **129**

5.1	Introduction	129
5.2	Brownian Motion with Drift	132

5.3   Brownian Motion as a Markov Process   132

5.4   Brownian Motion as a Martingale   133

5.5   First Passage Time of a Brownian Motion   133

5.6   Maximum of a Brownian Motion   135

5.7   First Passage Time in an Interval   135

5.8   The Brownian Bridge   136

5.9   Geometric Brownian Motion   137

5.10   The Langevin Equation   137

5.11   Summary   141

      Problems   141

6   INTRODUCTION TO STOCHASTIC CALCULUS   143

6.1   Introduction   143

6.2   The Ito Integral   145

6.3   The Stochastic Differential   146

6.4   The Ito's Formula   147

6.5   Stochastic Differential Equations   147

6.6   Solution of the Geometric Brownian Motion   148

6.7   The Ornstein–Uhlenbeck Process   151

    6.7.1   Solution of the Ornstein–Uhlenbeck SDE   152

    6.7.2   First Alternative Solution Method   153

    6.7.3   Second Alternative Solution Method   154

6.8   Mean-Reverting Ornstein–Uhlenbeck Process   155

6.9   Summary   157

7   DIFFUSION PROCESSES   158

7.1   Introduction   158

7.2   Mathematical Preliminaries   159

7.3   Diffusion on One-Dimensional Random Walk   160

    7.3.1   Alternative Derivation   163

7.4   Examples of Diffusion Processes   164

    7.4.1   Brownian Motion   164

    7.4.2   Brownian Motion with Drift   167

7.5   Correlated Random Walk and the Telegraph Equation   167

7.6   Diffusion at Finite Speed   170

7.7   Diffusion on Symmetric Two-Dimensional Lattice
     Random Walk   171

7.8   Diffusion Approximation of the Pearson Random Walk   173

7.9   Summary   174

8 LEVY WALK **175**

8.1 Introduction 175
8.2 Generalized Central Limit Theorem 175
8.3 Stable Distribution 177
8.4 Self-Similarity 182
8.5 Fractals 183
8.6 Levy Distribution 185
8.7 Levy Process 186
8.8 Infinite Divisibility 186
    8.8.1 The Infinite Divisibility of the Poisson Process 187
    8.8.2 Infinite Divisibility of the Compound Poisson Process 187
    8.8.3 Infinite Divisibility of the Brownian Motion with Drift 188
8.9 Levy Flight 188
    8.9.1 First Passage Time of Levy Flights 190
    8.9.2 Leapover Properties of Levy Flights 190
8.10 Truncated Levy Flight 191
8.11 Levy Walk 191
    8.11.1 Levy Walk as a Coupled CTRW 192
    8.11.2 Truncated Levy Walk 195
8.12 Summary 195

9 FRACTIONAL CALCULUS AND ITS APPLICATIONS **196**

9.1 Introduction 196
9.2 Gamma Function 197
9.3 Mittag–Leffler Functions 198
9.4 Laplace Transform 200
9.5 Fractional Derivatives 202
9.6 Fractional Integrals 203
9.7 Definitions of Fractional Integro-Differentials 203
    9.7.1 Riemann–Liouville Fractional Derivative 204
    9.7.2 Caputo Fractional Derivative 205
    9.7.3 Grunwald–Letnikov Fractional Derivative 206
9.8 Fractional Differential Equations 207
    9.8.1 Relaxation Differential Equation of Integer Order 208
    9.8.2 Oscillation Differential Equation of Integer Order 208
    9.8.3 Relaxation and Oscillation Fractional Differential Equations 209

9.9   Applications of Fractional Calculus   210

    9.9.1   Fractional Brownian Motion   210

    9.9.2   Multifractional Brownian Motion   213

    9.9.3   Fractional Random Walk   213

    9.9.4   Fractional (or Anomalous) Diffusion   215

    9.9.5   Fractional Gaussian Noise   221

    9.9.6   Fractional Poisson Process   222

9.10   Summary   224

10 PERCOLATION THEORY   225

10.1   Introduction   225

10.2   Graph Theory Revisited   226

    10.2.1   Complete Graphs   226

    10.2.2   Random Graphs   226

10.3   Percolation on a Lattice   228

    10.3.1   Cluster Formation and Phase Transition   229

    10.3.2   Percolation Probability and Critical Exponents   233

10.4   Continuum Percolation   235

    10.4.1   The Boolean Model   235

    10.4.2   The Random Connection Model   236

10.5   Bootstrap (or k-Core) Percolation   237

10.6   Diffusion Percolation   237

    10.6.1   Bootstrap Percolation versus Diffusion Percolation   239

10.7   First-Passage Percolation   239

10.8   Explosive Percolation   240

10.9   Percolation in Complex Networks   242

    10.9.1   Average Path Length   243

    10.9.2   Clustering Coefficient   243

    10.9.3   Degree Distribution   244

    10.9.4   Percolation and Network Resilience   244

10.10 Summary   245

REFERENCES   247

INDEX   253

PREFACE

This book is about random walks and the related subject of diffusion processes. The simplest definition of a random walk is that it is a stochastic process that consists of a sequence of discrete steps of fixed length. A more rigorous mathematical definition of a random walk is that it is a stochastic process that is formed by successive summation of independent and identically distributed random variables. In other words, it is a mathematical formalization of a trajectory that consists of taking successive random steps.

Random walk is related to numerous physical processes, including the Brownian motion, diffusion processes, and Levy flights. Consequently, it is used to explain observed behaviors of processes in several fields, including ecology, economics, psychology, computer science, physics, chemistry, and biology. The path traced by a molecule as it travels in a liquid or a gas, the search path of a foraging animal, the price of a fluctuating stock, and the financial status of a gambler can all be modeled as random walks. For example, the trajectory of insects on a horizontal planar surface may be accurately modeled as a random walk. In polymer physics, random walk is the simplest model to study polymers. Random walk accurately explains the relation between the time needed to make a decision and the probability that a certain decision will be made, which is a concept that is used in psychology. It is used in wireless networking to model node movements and node failures. It has been used in computer science to estimate the size of the World Wide Web. Also, in image processing, the random walker segmentation algorithm is used to determine the labels to associate with each pixel. Thus, random walk is a stochastic process that has proven to be a useful model in understanding several processes across a wide spectrum of scientific disciplines.

There are different types of random walks. Some random walks are on graphs while others are on the line, in the plane, or in higher dimensions. Random walks also vary with regard to the time parameter. In discrete-time random walks, the walker usually takes fixed-length steps in discrete time. In continuous-time random walks, the walker takes steps at random times, and the step length is usually a random variable.

The purpose of this book is to bring into one volume the different types of random walks and related topics, including Brownian motion, diffusion processes, and Levy flights. While many books have been written on random walks, our approach in this book is different. The book includes standard methods and applications of Brownian motion, which is considered a limit of random walk, and a discussion on Levy flights and Levy walk, which have become popular in random searches in ecology, finance, and other fields. It also includes a chapter on fractional calculus that is necessary for understanding anomalous (or fractional) diffusion, fractional Brownian motion, and fractional random walk. Finally, it introduces the reader to percolation theory and its relationship to diffusion processes. It is a self-contained book that will appeal to graduate students across science, engineering, and mathematics who need to understand the applications of random walk and diffusion process techniques, as well as to established researchers. It presents the connections between diffusion equations and random walks and introduces stochastic calculus, which is a prerequisite for understanding some of the modern concepts in Brownian motion and diffusion processes.

The chapters are organized as follows. Chapter 1 presents an introduction to probability while Chapter 2 gives an overview of stochastic processes. Chapter 3 discusses one-dimensional random walk while Chapter 4 discusses two-dimensional random walk. Chapter 5 discusses Brownian motion, Chapter 6 presents an introduction to stochastic calculus, and Chapter 7 discusses diffusion processes. Chapter 8 discusses Levy flights and Levy walk, Chapter 9 discusses fractional calculus and its applications, and Chapter 10 discusses percolation theory.

The book is written with the understanding that much of research on social networks, economics, finance, ecology, biostatistics, polymer physics, and population genetics has become interdisciplinary with random walk and diffusion processes as the common thread. Thus, it is truly a book designed for interdisciplinary use. It can be used for a one-semester course on stochastic processes and their applications.

OLIVER C. IBE

ACKNOWLEDGMENTS

My journey into the field of stochastic systems modeling started with my encounter at the Massachusetts Institute of Technology with two giants in the field, namely, the late Professor Alvin Drake, who was my academic adviser and under whom I was a teaching assistant in a course titled "Probabilistic Systems Analysis," and Professor Robert Gallager, who was my adviser for both my master's and doctoral theses. This book is a product of the wonderful instruction I received from these two great professors, and I sincerely appreciate all that they did to get me excited in this field.

This is the second project that I have completed with my editor, Ms. Susanne Steitz-Filler of Wiley. I am sincerely grateful to her for encouraging me and for being patient with me throughout the time it took to get the project completed. I would also like to thank Ms. Sari Friedman, an Editorial Assistant at Wiley, for ensuring that the production schedule is met. This is also my second project with her. I am grateful to the anonymous reviewers for their comments that helped to improve the quality of the book.

Finally, I would like to thank my wife, Christina, and our children Chidinma, Ogechi, Amanze, and Ugonna, for the joy they have brought to my life. I could not have completed this project without their encouragement.

CHAPTER 1

REVIEW OF PROBABILITY THEORY

1.1 INTRODUCTION

The concepts of *experiments* and *events* are very important in the study of probability. In probability, an experiment is any process of trial and observation. An experiment whose outcome is uncertain before it is performed is called a *random* experiment. When we perform a random experiment, the collection of possible elementary outcomes is called the *sample space* of the experiment, which is usually denoted by Ω. We define these outcomes as elementary outcomes because exactly one of the outcomes occurs when the experiment is performed. The elementary outcomes of an experiment are called the *sample points* of the sample space and are denoted by w_i, $i=1, 2, \ldots$. If there are n possible outcomes of an experiment, then the sample space is $\Omega=\{w_1, w_2, \ldots, w_n\}$. An *event* is the occurrence of either a prescribed outcome or any one of a number of possible outcomes of an experiment. Thus, an event is a subset of the sample space.

1.2 RANDOM VARIABLES

Consider a random experiment with sample space Ω. Let w be a sample point in Ω. We are interested in assigning a real number to each $w \in \Omega$. A random variable, $X(w)$, is a single-valued real function that assigns a real number, called the value of $X(w)$, to each sample point $w \in \Omega$. That is, it is a mapping of the sample space onto the real line.

Elements of Random Walk and Diffusion Processes, First Edition. Oliver C. Ibe.
© 2013 John Wiley & Sons, Inc. Published 2013 by John Wiley & Sons, Inc.

Generally, a random variable is represented by a single letter X instead of the function $X(w)$. Therefore, in the remainder of the book we use X to denote a random variable. The sample space Ω is called the *domain* of the random variable X. Also, the collection of all numbers that are values of X is called the *range* of the random variable X.

Let X be a random variable and x a fixed real value. Let the event A_x define the subset of Ω that consists of all real sample points to which the random variable X assigns the number x. That is,

$$A_x = \{w \mid X(w) = x\} = [X = x]$$

Since A_x is an event, it will have a probability, which we define as follows:

$$p = P[A_x]$$

We can define other types of events in terms of a random variable. For fixed numbers x, a, and b, we can define the following:

$$[X \leq x] = \{w \mid X(w) \leq x\}$$
$$[X > x] = \{w \mid X(w) > x\}$$
$$[a < X < x] = \{w \mid a < X(w) < b\}$$

These events have probabilities that are denoted by the following:

- $P[X \leq x]$ is the probability that X takes a value less than or equal to x.
- $P[X > x]$ is the probability that X takes a value greater than x; this is equal to $1 - P[X \leq x]$.
- $P[a < X < b]$ is the probability that X takes a value that strictly lies between a and b.

1.2.1   Distribution Functions

Let X be a random variable and x be a number. As stated earlier, we can define the event $[X \leq x] = \{x \mid X(w) \leq x\}$. The distribution function (or the cumulative distribution function (CDF)) of X is defined by

$$F_X(x) = P[X \leq x], \quad -\infty < x < \infty \tag{1.1}$$

That is, $F_X(x)$ denotes the probability that the random variable X takes on a value that is less than or equal to x. Some properties of $F_X(x)$ include

1. $F_X(x)$ is a nondecreasing function, which means that if $x_1 < x_2$, then $F_X(x_1) \leq F_X(x_2)$. Thus, $F_X(x)$ can increase or stay level, but it cannot go down.
2. $0 \leq F_X(x) \leq 1$
3. $F_X(\infty) = 1$
4. $F_X(-\infty) = 0$

5. $P[a<X\le b]=F_X(b)-F_X(a)$
6. $P[X>a]=1-P[X\le a]=1-F_X(a)$

1.2.2   Discrete Random Variables

A discrete random variable is a random variable that can take on at most a countable number of possible values. For a discrete random variable X, the *probability mass function* (PMF), $p_X(x)$, is defined as follows:

$$p_X(x) = P[X = x] \tag{1.2}$$

The PMF is nonzero for at most a countable or countably infinite number of values of x. In particular, if we assume that X can only assume one of the values $x_1, x_2, \ldots, x_n$, then

$$p_X(x_i) \ge 0 \quad i = 1, 2, \ldots, n$$
$$p_X(x) = 0 \quad \text{otherwise}$$

The CDF of X can be expressed in terms of $p_X(x)$ as follows:

$$F_X(x) = \sum_{k \le x} p_X(k) \tag{1.3}$$

The CDF of a discrete random variable is a step function. That is, if X takes on values $x_1, x_2, x_3, \ldots$, where $x_1 < x_2 < x_3, \ldots$, then the value of $F_X(x)$ is constant in the interval between x_{i-1} and x_i and then takes a jump of size $p_X(x_i)$ at x_i, $i=2, 3, \ldots$. Thus, in this case, $F_X(x)$ represents the sum of all the probability masses we have encountered as we move from $-\infty$ to x.

1.2.3   Continuous Random Variables

Discrete random variables have a set of possible values that are either finite or countably infinite. However, there exists another group of random variables that can assume an uncountable set of possible values. Such random variables are called continuous random variables. Thus, we define a random variable X to be a continuous random variable if there exists a nonnegative function $f_X(x)$, defined for all real $x \in (-\infty, \infty)$, having the property that for any set A of real numbers,

$$P[X \in A] = \int_A f_X(x)dx \tag{1.4}$$

The function $f_X(x)$ is called the *probability density function* (PDF) of the random variable X and is defined by

$$f_X(x) = \frac{dF_X(x)}{dx} \tag{1.5}$$

The properties of $f_X(x)$ are as follows:

1. $f_X(x) \geq 0$
2. Since X must assume some value, $\int_{-\infty}^{\infty} f_X(x)dx = 1$
3. $P[a \leq X \leq b] = \int_a^b f_X(x)dx$, which means that $P[X = a] = \int_a^a f_X(x)dx = 0$. Thus, the probability that a continuous random variable will assume any fixed value is 0.
4. $P[X < a] = P[X \leq a] = \int_{-\infty}^a f_X(x)dx = F_X(a)$

1.2.4 Expectations

If X is a random variable, then the *expectation* (or *expected value* or *mean*) of X, denoted by $E[X]$, is defined by

$$E[X] = \begin{cases} \sum_i x_i p_X(x_i) & X \text{ discrete} \\ \int_{-\infty}^{\infty} x f_X(x)dx & X \text{ continuous} \end{cases} \tag{1.6}$$

Thus, the expected value of X is a weighted average of the possible values that X can take, where each value is weighted by the probability that X takes that value. The expected value of X is sometimes denoted by $\overline{X}$ and by $\langle X \rangle$.

1.2.5 Moments of Random Variables and the Variance

The nth moment of the random variable X, denoted by $E[X^n] = \overline{X^n}$, is defined by

$$E[X^n] = \overline{X^n} = \begin{cases} \sum_i x_i^n p_X(x_i) & X \text{ discrete} \\ \int_{-\infty}^{\infty} x^n f_X(x)dx & X \text{ continuous} \end{cases} \tag{1.7}$$

for $n = 1, 2, 3, \ldots$. The first moment, $E[X]$, is the expected value of X.

We can also define the *central moments* (or *moments about the mean*) of a random variable. These are the moments of the difference between a random variable and its expected value. The nth central moment is defined by

$$E[(X - \overline{X})^n] = \overline{(X - \overline{X})^n} = \begin{cases} \sum_i (x_i - \overline{X})^n p_X(x_i) & X \text{ discrete} \\ \int_{-\infty}^{\infty} (x - \overline{X})^n f_X(x)dx & X \text{ continuous} \end{cases} \tag{1.8}$$

The central moment for the case of $n = 2$ is very important and carries a special name, the *variance*, which is usually denoted by σ_X^2. Thus,

$$\sigma_X^2 = E\left[\left(X - \overline{X}\right)^2\right] = \overline{\left(X - \overline{X}\right)^2} = \begin{cases} \sum_i \left(x_i - \overline{X}\right)^2 p_X(x_i) & X \text{ discrete} \\ \int_{-\infty}^{\infty} \left(x - \overline{X}\right)^2 f_X(x)dx & X \text{ continuous} \end{cases} \quad (1.9)$$

It can be shown that

$$\sigma_X^2 = E[X^2] - \{E[X]\}^2 \tag{1.10}$$

1.3 TRANSFORM METHODS

Different types of transforms are used in science and engineering. These include the z-transform, Laplace transform, and Fourier transform. One of the reasons for their popularity is that when they are introduced into the solution of many problems, the calculations become greatly simplified. For example, many solutions of equations that involve derivatives and integrals of functions are given as the convolution of two functions: $a(x) * b(x)$. As students of signal and systems know, the Fourier transform of a convolution is the product of the individual Fourier transforms. That is, if $F[g(x)]$ is the Fourier transform of the function $g(x)$, then

$$F[a(x) * b(x)] = A(w)B(w)$$

where $A(w)$ is the Fourier transform of $a(x)$ and $B(w)$ is the Fourier transform of $b(x)$. This means that the convolution operation can be replaced by the much simpler multiplication operation. In fact, sometimes transform methods are the only tools available for solving some types of problems.

We consider three types of transforms: the characteristic function of PDFs, the z-transform (or moment generating function) of PMFs, and the s-transform (or Laplace transform) of PDFs. The s-transform and the z-transform are particularly used when random variables take only nonnegative values, which is usually the case in many applications discussed in this book.

1.3.1 The Characteristic Function

Let $f_X(x)$ be the PDF of the continuous random variable X. The characteristic function of X is defined by

$$\Phi_X(w) = E[e^{jwX}] = \int_{-\infty}^{\infty} e^{jwx} f_X(x)dx \tag{1.11}$$

where $j = \sqrt{-1}$. We can obtain $f_X(x)$ from $\Phi_X(w)$ as follows:

$$f_X(x) = \frac{1}{2\pi} \int_{-\infty}^{\infty} \Phi_X(w) e^{-jwx} dw \tag{1.12}$$

If X is a discrete random variable with PMF $p_X(x)$, the characteristic function is given by

$$\Phi_X(w) = \sum_{x=-\infty}^{\infty} p_X(x) e^{jwx} \tag{1.13}$$

Note that $\Phi_X(0)=1$, which is a test of whether a given function of w is a true characteristic function of the PDF or PMF of a random variable.

1.3.2 Moment-Generating Property of the Characteristic Function

One of the primary reasons for studying the transform methods is to use them to derive the moments of the different probability distributions. By definition

$$\Phi_X(w) = \int_{-\infty}^{\infty} e^{jwx} f_X(x)\, dx$$

Taking the derivative of $\Phi_X(w)$, we obtain

$$\frac{d}{dw}\Phi_X(w) = \frac{d}{dw}\int_{-\infty}^{\infty} e^{jwx} f_X(x)dx = \int_{-\infty}^{\infty} \frac{d}{dw} e^{jwx} f_X(x)dx$$

$$= \int_{-\infty}^{\infty} jx e^{jwx} f_X(x)dx$$

$$\frac{d}{dw}\Phi_X(w)\Big|_{w=0} = \int_{-\infty}^{\infty} jx f_X(x)dx = jE[X]$$

$$\frac{d^2}{dw^2}\Phi_X(w) = \frac{d}{dw}\int_{-\infty}^{\infty} jx e^{jwx} f_X(x)dx = -\int_{-\infty}^{\infty} x^2 e^{jwx} f_X(x)dx$$

$$\frac{d^2}{dw^2}\Phi_X(w)\Big|_{w=0} = -\int_{-\infty}^{\infty} x^2 f_X(x)dx = -E[X^2]$$

In general,

$$\frac{d^n}{dw^n}\Phi_X(w)\Big|_{w=0} = j^n E[X^n] \tag{1.14}$$

1.3.3 The s-Transform

Let $f_X(x)$ be the PDF of the continuous random variable X that takes only nonnegative values; that is, $f_X(x)=0$ for $x<0$. The s-transform of $f_X(x)$, denoted by $M_X(s)$, is defined by

$$M_X(s) = E[e^{-sX}] = \int_0^{\infty} e^{-sx} f_X(x)\, dx \tag{1.15}$$

One important property of an s-transform is that when it is evaluated at the point $s=0$, its value is equal to 1. That is,

$$M_X(s)\Big|_{s=0} = \int_0^{\infty} f_X(x)\, dx = 1$$

For example, the value of K for which the function $A(s) = K/(s+5)$ is a valid s-transform of a PDF is obtained by setting $A(0) = 1$, which gives $K = 5$.

1.3.4 Moment-Generating Property of the s-Transform

As stated earlier, one of the primary reasons for studying the transform methods is to use them to derive the moments of the different probability distributions. By definition

$$M_X(s) = \int_0^\infty e^{-sx} f_X(x) dx$$

Taking different derivatives of $M_X(s)$ and evaluating them at $s = 0$, we obtain the following results:

$$\frac{d}{ds} M_X(s) = \frac{d}{ds} \int_0^\infty e^{-sx} f_X(x) dx = \int_0^\infty \frac{d}{ds} e^{-sx} f_X(x) dx$$

$$= -\int_0^\infty x e^{-sx} f_X(x) dx$$

$$\frac{d}{ds} M_X(s)\Big|_{s=0} = -\int_0^\infty x f_X(x) dx = -E[X]$$

$$\frac{d^2}{ds^2} M_X(s) = \frac{d}{ds}(-1)\int_0^\infty x e^{-sx} f_X(x) dx = \int_0^\infty x^2 e^{-sx} f_X(x) dx$$

$$\frac{d^2}{ds^2} M_X(s)\Big|_{s=0} = \int_0^\infty x^2 f_X(x) dx = E[X^2]$$

In general,

$$\frac{d^n}{ds^n} M_X(s)\Big|_{s=0} = (-1)^n E[X^n] \tag{1.16}$$

1.3.5 The z-Transform

Let $p_X(x)$ be the PMF of the nonnegative discrete random variable X. The z-transform of $p_X(x)$, denoted by $G_X(z)$, is defined by

$$G_X(z) = E[z^X] = \sum_{x=0}^\infty z^x p_X(x) \tag{1.17}$$

The sum is guaranteed to converge and, therefore, the z-transform exists, when evaluated on or within the unit circle (where $|z| \le 1$). Note that

$$G_X(1) = \sum_{x=0}^\infty p_X(x) = 1$$

This means that a valid z-transform of a PMF reduces to unity when evaluated at $z=1$. However, this is a necessary but not sufficient condition for a function to the z-transform of a PMF. By definition,

$$G_X(z) = \sum_{x=0}^{\infty} z^x p_X(x) = p_X(0) + z p_X(1) + z^2 p_X(2) + z^3 p_X(3) + \cdots$$

This means that $P[X=k]=p_X(k)$ is the coefficient of z^k in the series expansion. Thus, given the z-transform of a PMF, we can uniquely recover the PMF. The implication of this statement is that not every function of z that has a value 1 when evaluated at $z=1$ is a valid z-transform of a PMF. For example, consider the function $A(z)=2z-1$. Although $A(1)=1$, the function contains invalid coefficients in the sense that these coefficients either have negative values or positive values that are greater than one. Thus, for a function of z to be a valid z-transform of a PMF, it must have a value of 1 when evaluated at $z=1$, and the coefficients of z must be nonnegative numbers that cannot be greater than 1.

The individual terms of the PMF can also be determined as follows:

$$p_X(x) = \frac{1}{x!} \left[\frac{d^x}{dz^x} G_X(z) \right]_{z=0} \qquad x = 0, 1, 2, \ldots \qquad (1.18)$$

This feature of the z-transform is the reason it is sometimes called the *probability-generating function.*

1.3.6 Moment-Generating Property of the z-Transform

As stated earlier, one of the major motivations for studying transform methods is their usefulness in computing the moments of the different random variables. Unfortunately, the moment-generating capability of the z-transform is not as computationally efficient as that of the s-transform.

The moment-generating capability of the z-transform lies in the results obtained from evaluating the derivatives of the transform at $z=1$. For a discrete random variable X with PMF $p_X(x)$, we have that

$$G_X(z) = \sum_{x=0}^{\infty} z^x p_X(x)$$

$$\frac{d}{dz} G_X(z) = \frac{d}{dz} \sum_{x=0}^{\infty} z^x p_X(x) = \sum_{x=0}^{\infty} \frac{d}{dz} z^x p_X(x) = \sum_{x=0}^{\infty} x z^{x-1} p_X(x)$$

$$= \sum_{x=1}^{\infty} x z^{x-1} p_X(x)$$

$$\frac{d}{dz} G_X(z) \bigg|_{z=1} = \sum_{x=1}^{\infty} x p_X(x) = \sum_{x=0}^{\infty} x p_X(x) = E[X]$$

Similarly,

$$\frac{d^2}{dz^2} G_X(z) = \frac{d}{dz} \sum_{x=1}^{\infty} xz^{x-1} p_X(x) = \sum_{x=1}^{\infty} x \frac{d}{dz} z^{x-1} p_X(x)$$

$$= \sum_{x=1}^{\infty} x(x-1) z^{x-2} p_X(x)$$

$$\frac{d^2}{dz^2} G_X(z)|_{z=1} = \sum_{x=1}^{\infty} x(x-1) p_X(x) = \sum_{x=0}^{\infty} x(x-1) p_X(x)$$

$$= \sum_{x=0}^{\infty} x^2 p_X(x) - \sum_{x=0}^{\infty} x p_X(x)$$

$$= E[X^2] - E[X]$$

$$E[X^2] = \frac{d^2}{dz^2} G_X(z)|_{z=1} + \frac{d}{dz} G_X(z)|_{z=1}$$

Thus, the variance is obtained as follows:

$$\sigma_X^2 = E[X^2] - \{E[X]\}^2 = \left[\frac{d^2}{dz^2} G_X(z) + \frac{d}{dz} G_X(z) - \left\{ \frac{d}{dz} G_X(z) \right\}^2 \right]_{z=1}$$

1.4 COVARIANCE AND CORRELATION COEFFICIENT

Consider two random variables X and Y with expected values $E[X] = \mu_X$ and $E[Y] = \mu_Y$, respectively, and variances σ_X^2 and σ_Y^2, respectively. The *covariance* of X and Y, which is denoted by $\text{Cov}(X, Y)$ or σ_{XY}, is defined by

$$\begin{aligned} \text{Cov}(X, Y) = \sigma_{XY} &= E[(X - \mu_X)(Y - \mu_Y)] \\ &= E[XY - \mu_Y X - \mu_X Y + \mu_X \mu_Y] \\ &= E[XY] - \mu_X \mu_Y \end{aligned} \quad (1.19)$$

If X and Y are independent, then $E[XY] = E[X]E[Y] = \mu_X \mu_Y$ and $\text{Cov}(X, Y) = 0$. However, the converse is not true; that is, if the covariance of X and Y is 0, it does not necessarily mean that X and Y are independent random variables. If the covariance of two random variables is 0, we define the two random variables to be *uncorrelated*.

We define the *correlation coefficient* of X and Y, denoted by $\rho(X, Y)$ or ρ_{XY}, as follows:

$$\rho_{XY} = \frac{\text{Cov}(X, Y)}{\sqrt{\sigma_X^2 \sigma_Y^2}} = \frac{\sigma_{XY}}{\sigma_X \sigma_Y} \quad (1.20)$$

The correlation coefficient has the property that

$$-1 \le \rho_{XY} \le 1 \quad (1.21)$$

1.5   SUMS OF INDEPENDENT RANDOM VARIABLES

Consider two independent continuous random variables X and Y. We are interested in computing the CDF and PDF of their sum $g(X, Y) = U = X + Y$. The random variable U can be used to model the reliability of systems with standby connections. In such systems, the component A whose time-to-failure is represented by the random variable X is the primary component, and the component B whose time-to-failure is represented by the random variable Y is the backup component that is brought into operation when the primary component fails. Thus, U represents the time until the system fails, which is the sum of the lifetimes of both components.

Their CDF can be obtained as follows:

$$F_U(u) = P[U \le u] = P[X + Y \le u] = \iint_D f_{XY}(x, y)dx\, dy$$

where $f_{XY}(x, y)$ is the joint PDF of X and Y and D is the set $D = \{(x, y) | x + y \le u\}$.

Thus,

$$F_U(u) = \int_{-\infty}^{\infty}\int_{-\infty}^{u-y} f_{XY}(x, y)dx\, dy = \int_{-\infty}^{\infty}\int_{-\infty}^{u-y} f_X(x)f_Y(y)dx\, dy$$

$$= \int_{-\infty}^{\infty} \left\{ \int_{-\infty}^{u-y} f_X(x)dx \right\} f_Y(y)dy = \int_{-\infty}^{\infty} F_X(u - y)f_Y(y)dy$$

The PDF of U is obtained by differentiating the CDF, as follows:

$$f_U(u) = \frac{d}{du} F_U(u) = \frac{d}{du}\int_{-\infty}^{\infty} F_X(u - y)f_Y(y)dy = \int_{-\infty}^{\infty} \frac{d}{du} F_X(u - y)f_Y(y)dy$$

$$= \int_{-\infty}^{\infty} f_X(u - y)f_Y(y)dy$$

where we have assumed that we can interchange differentiation and integration. The last equation is a well-known result in signal analysis called the *convolution integral*. Thus, we find that the PDF of the sum U of two independent random variables X and Y is the convolution of the PDFs of the two random variables; that is,

$$f_U(u) = f_X(u) * f_Y(u) \tag{1.22}$$

In general, if U is the sum on n mutually independent random variables $X_1, X_2, \ldots, X_n$ whose PDFs are $f_{X_i}(x)$, $i = 1, 2, \ldots, n$, then we have that

$$U = X_1 + X_2 + \cdots + X_n$$
$$f_U(u) = f_{X_1}(u) * f_{X_2}(u) * \cdots * f_{X_n}(u) \tag{1.23}$$

Thus, the s-transform of the PDF of U is given by

$$M_U(s) = \prod_{i=1}^{n} M_{X_i}(s) \tag{1.24}$$

1.6 SOME PROBABILITY DISTRIBUTIONS

Random variables with special probability distributions are encountered in different fields of science and engineering. In this section, we describe some of these distributions, including their expected values, variances, and s-transforms (or z-transforms or characteristic functions, as the case may be).

1.6.1 The Bernoulli Distribution

A Bernoulli trial is an experiment that results in two outcomes: *success* and *failure*. One example of a Bernoulli trial is the coin-tossing experiment, which results in heads or tails. In a Bernoulli trial, we define the probability of success and probability of failure as follows:

$$P[\text{success}] = p \quad 0 \le p \le 1$$
$$P[\text{failure}] = 1 - p$$

Let us associate the events of the Bernoulli trial with a random variable X such that when the outcome of the trial is a success we define $X=1$, and when the outcome is a failure we define $X=0$. The random variable X is called a Bernoulli random variable, and its PMF is given by

$$p_X(x) = \begin{cases} 1-p & x = 0 \\ p & x = 1 \end{cases} \tag{1.25}$$

An alternative way to define the PMF of X is as follows:

$$p_X(x) = p^x(1-p)^{1-x} \quad x = 0,1 \tag{1.26}$$

The CDF is given by

$$F_X(x) = \begin{cases} 0 & x < 0 \\ 1-p & 0 \le x < 1 \\ 1 & x \ge 1 \end{cases} \tag{1.27}$$

The expected value of X is given by

$$E[X] = 0(1-p) + 1(p) = p \tag{1.28}$$

Similarly, the second moment of X is given by

$$E[X^2] = 0^2(1-p) + 1^2(p) = p$$

Thus, the variance of X is given by

$$\sigma_X^2 = E[X^2] - \{E[X]\}^2 = p - p^2 = p(1-p) \tag{1.29}$$

The z-transform of the PMF is given by

$$G_X(z) = \sum_{x=0}^{\infty} z^x p_X(x) = z^0 p_X(0) + z^1 p_X(1) = 1 - p + zp \tag{1.30}$$

1.6.2 The Binomial Distribution

Suppose we conduct n independent Bernoulli trials and we represent the number of successes in those n trials by the random variable $X(n)$. Then $X(n)$ is defined as a binomial random variable with parameters (n, p). The PMF of a random variable $X(n)$ with parameters (n, p) is given by

$$p_{X(n)}(x) = \binom{n}{x} p^x (1-p)^{n-x} \quad x = 0, 1, 2, \ldots, n \tag{1.31}$$

The binomial coefficient, $\binom{n}{x}$, represents the number of ways of arranging x successes and $n-x$ failures.

The CDF, mean, and variance of $X(n)$ and the z-transform of its PMF are given by

$$F_{X(n)}(x) = P[X(n) \le x] = \sum_{k=0}^{x} \binom{n}{k} p^k (1-p)^{n-k} \tag{1.32a}$$

$$E[X(n)] = np \tag{1.32b}$$

$$\sigma_{X(n)}^2 = np(1-p) \tag{1.32c}$$

$$G_{X(n)}(z) = (1 - p + zp)^n \tag{1.32d}$$

1.6.3 The Geometric Distribution

The geometric random variable is used to describe the number of Bernoulli trials until the first success occurs. Let X be a random variable that denotes the number of Bernoulli trials until the first success. If the first success occurs on the xth trial, then we know that the first $x-1$ trials resulted in failures. Thus, the PMF of a geometric random variable, X, is given by

$$p_X(x) = p(1-p)^{x-1} \quad x = 1, 2, \ldots \tag{1.33}$$

The CDF, mean, and variance of X and the z-transform of its PMF are given by

$$F_X(x) = 1 - (1-p)^x \tag{1.34a}$$

$$E[X] = \frac{1}{p} \qquad (1.34\text{b})$$

$$\sigma_X^2 = \frac{1-p}{p} \qquad (1.34\text{c})$$

$$G_X(z) = \frac{zp}{1 - z(1-p)} \qquad (1.34\text{d})$$

1.6.4   The Poisson Distribution

A discrete random variable K is called a Poisson random variable with parameter λ, where $\lambda > 0$, if its PMF is given by

$$p_K(k) = \frac{\lambda^k}{k!} e^{-\lambda} \quad k = 0, 1, 2, \dots \qquad (1.35)$$

The CDF, mean, and variance of K and the z-transform of its PMF are given by

$$F_K(k) = \sum_{r=0}^{k} \frac{\lambda^r}{r!} e^{-\lambda} \qquad (1.36\text{a})$$

$$E[K] = \lambda \qquad (1.36\text{b})$$

$$\sigma_K^2 = \lambda \qquad (1.36\text{c})$$

$$G_K(z) = e^{-\lambda(1-z)} \qquad (1.36\text{d})$$

1.6.5   The Exponential Distribution

A continuous random variable X is defined to be an exponential random variable (or X has an exponential distribution) if for some parameter $\lambda > 0$ its PDF is given by

$$f_X(x) = \lambda e^{-\lambda x}, \quad x \geq 0 \qquad (1.37)$$

The CDF, mean, and variance of X and the s-transform of its PDF are given by

$$F_X(x) = 1 - e^{-\lambda x}, \quad x \geq 0 \qquad (1.38\text{a})$$

$$E[X] = \frac{1}{\lambda} \qquad (1.38\text{b})$$

$$\sigma_X^2 = \frac{1}{\lambda^2} \qquad (1.38\text{c})$$

$$M_X(s) = \frac{\lambda}{s + \lambda} \qquad (1.38\text{d})$$

Assume that X_1, X_2, ..., X_n is a set of independent and identically distributed exponential random variables with mean $E[X_i]=1/\lambda$. Let $X=X_1+X_2+\cdots+X_n$. Then X is defined as the nth-order *Erlang random variable*. One of the features of the exponential distribution is its *forgetfulness* property. Specifically,

$$P[X \le s+t \mid X > t] = P[X \le s] \Rightarrow f_{X|X>t}(x \mid X > t) = \lambda e^{-\lambda(x-t)}$$

1.6.6 Normal Distribution

A continuous random variable X is defined to be a normal random variable with parameters μ_X and σ_X^2 if its PDF is given by

$$f_X(x) = \frac{1}{\sqrt{2\pi\sigma_X^2}} e^{-(x-\mu_X)^2/2\sigma_X^2} \qquad -\infty < x < \infty \tag{1.39}$$

The PDF is a bell-shaped curve that is symmetric about μ_X, which is the mean of X. The parameter σ_X^2 is the variance. The CDF of X is given by

$$F_X(x) = P[X \le x] = \frac{1}{\sigma_X \sqrt{2\pi}} \int_{-\infty}^{x} e^{-(u-\mu_X)^2/2\sigma_X^2} du$$

The normal random variable X with parameters μ_X and σ_X^2 is usually designated $X \sim N(\mu_X, \sigma_X^2)$. The special case of zero mean and unit variance (i.e., $\mu_X=0$ and $\sigma_X^2=1$) is designated $X \sim N(0, 1)$ and is called the *standard normal random variable*. Let $y=(u-\mu_X)/\sigma_X$. Then, $du=\sigma_X dy$ and the CDF of X becomes

$$F_X(x) = \frac{1}{\sqrt{2\pi}} \int_{..}^{(x-\mu_X)/\sigma_X} e^{-y^2/2} dy$$

Thus, with the aforementioned transformation, X becomes a standard normal random variable. The aforementioned integral cannot be evaluated in closed form. It is usually evaluated numerically through the function $\Phi(x)$, which is defined as follows:

$$\Phi(x) = \frac{1}{\sqrt{2\pi}} \int_{-\infty}^{x} e^{-y^2/2} dy \tag{1.40}$$

Thus, the CDF of X is given by

$$F_X(x) = \frac{1}{\sqrt{2\pi}} \int_{-\infty}^{(x-\mu_X)/\sigma_X} e^{-y^2/2} dy = \Phi\left(\frac{x-\mu_x}{\sigma_X}\right) \tag{1.41}$$

The values of $\Phi(x)$ are usually given for nonnegative values of x. For negative values of x, $\Phi(x)$ can be obtained from the following relationship:

$$\Phi(-x) = 1 - \Phi(x) \tag{1.42}$$

Values of $\Phi(x)$ are given in standard books on probability, such as Ibe (2005). The characteristic function of $f_X(x)$ is obtained as follows:

$$\Phi_X(w) = \int_{-\infty}^{\infty} e^{jwx} f_X(x) dx = \frac{1}{\sqrt{2\pi\sigma_X^2}} \int_{-\infty}^{\infty} e^{jwx} e^{-(x-\mu_X)^2/2\sigma_X^2} dx$$

Let $u = (x - \mu_X)/\sigma_X$, which means that $x = u\sigma_X + \mu_X$ and $dx = \sigma_X du$. Thus,

$$\Phi_X(w) = \frac{1}{\sqrt{2\pi}} \int_{-\infty}^{\infty} e^{-u^2/2} e^{jw(u\sigma_X + \mu_X)} dx = \frac{e^{jw\mu_X}}{\sqrt{2\pi}} \int_{-\infty}^{\infty} e^{-(u^2 - 2jwu\sigma_X)/2} dx$$

Now,

$$\frac{u^2 - 2jwu\sigma_X}{2} = \frac{u^2 - 2jwu\sigma_X + (jw\sigma_X)^2}{2} - \frac{(jw\sigma_X)^2}{2}$$

$$= \frac{(u - jw\sigma_X)^2}{2} + \frac{w^2\sigma_X^2}{2}$$

Thus,

$$\Phi_X(w) = \frac{e^{jw\mu_X}}{\sqrt{2\pi}} \int_{-\infty}^{\infty} e^{-(u^2 - 2jwu\sigma_X)/2} du = \frac{e^{jw\mu_X}}{\sqrt{2\pi}} \int_{-\infty}^{\infty} e^{-(u - jw\sigma_X)^2/2} e^{-w^2\sigma_X^2/2} du$$

$$= e^{(jw\mu_X - w^2\sigma_X^2/2)} \left\{ \frac{1}{\sqrt{2\pi}} \int_{-\infty}^{\infty} e^{-(u - jw\sigma_X)^2/2} du \right\}$$

If we substitute $v = u - jw\sigma_X$, then the term in parentheses becomes

$$\frac{1}{\sqrt{2\pi}} \int_{-\infty}^{\infty} e^{-v^2/2} dv = 1$$

because the random variable $V \sim N(0, 1)$. Therefore,

$$\Phi_X(w) = e^{(jw\mu_X - w^2\sigma_X^2/2)} = \exp\left(jw\mu_X - \frac{w^2\sigma_X^2}{2} \right) \tag{1.43}$$

We state the following theorem whose proof can be found in any standard book on probability theory:

Theorem 1.1: Let $X_1, X_2, \ldots, X_n$ be independent normally distributed random variables with the mean and variance of X_i given by μ_i and σ_i^2, respectively. Then, the random variable

$$X = \sum_{i=1}^{n} X_i$$

has a normal distribution with mean and variance, respectively, given by

$$\mu = \sum_{i=1}^{n} \mu_i$$

$$\sigma^2 = \sum_{i=1}^{n} \sigma_i^2$$

That is, if $X_i \sim N(\mu_i, \sigma_i^2)$, $i = 1, 2, \ldots, n$, then $X \sim N(\mu, \sigma^2)$. For example, let X_1 and X_2 be independent and identically distributed standard random variables. We can compute $P[-1 \le X_1 + X_2 \le 3]$ by noting that according to the preceding theorem, $X = X_1 + X_2 \sim N(0, 2)$. Thus, if $F_X(x)$ is the CDF of X, we have that

$$P[-1 \le X_1 + X_2 \le 3] = P[-1 \le X \le 3] = F_X(3) - F_X(-1)$$

$$= \Phi\left(\frac{3-0}{\sqrt{2}}\right) - \Phi\left(\frac{-1-0}{\sqrt{2}}\right) = \Phi\left(\frac{3\sqrt{2}}{2}\right) - \Phi\left(\frac{-\sqrt{2}}{2}\right)$$

$$= \Phi(2.12) - \Phi(-0.71) = \Phi(2.12) - \{1 - \Phi(0.71)\}$$

$$= \Phi(2.12) + \Phi(0.71) - 1 = 0.9830 + 0.7611 - 1$$

$$= 0.7441$$

1.7   LIMIT THEOREMS

In this section, we discuss two fundamental theorems in probability. These are the law of large numbers, which is regarded as the first fundamental theorem, and the central limit theorem, which is regarded as the second fundamental theorem. We begin the discussion with the Markov and Chebyshev inequalities that enable us to prove these theorems.

1.7.1   Markov Inequality

The Markov inequality applies to random variables that take only nonnegative values. It can be stated as follows:

Proposition 1.1: If X is a random variable that takes only nonnegative values, then for any $a > 0$,

$$P[X \ge a] \le \frac{E[X]}{a} \tag{1.44}$$

Proof: We consider only the case when X is a continuous random variable. Thus,

$$E[X] = \int_0^{\infty} x f_X(x) dx = \int_0^{a} x f_X(x) dx + \int_a^{\infty} x f_X(x) dx \ge \int_a^{\infty} x f_X(x) dx$$

$$\ge \int_a^{\infty} a f_X(x) dx = a \int_a^{\infty} f_X(x) dx = a P[X \ge a]$$

and the result follows.

1.7.2 Chebyshev Inequality

The Chebyshev inequality enables us to obtain bounds on probability when both the mean and variance of a random variable are known. The inequality can be stated as follows:

Proposition 1.2: Let X be a random variable with mean μ and variance σ^2. Then, for any $b>0$,

$$P\big[\,|\,X-\mu\,|\geq b\,\big]\leq \frac{\sigma^2}{b^2} \tag{1.45}$$

Proof: Since $(X-\mu)^2$ is a nonnegative random variable, we can invoke the Markov inequality, with $a=b^2$, to obtain

$$P\big[(X-\mu)^2\geq b^2\big]\leq \frac{E\big[(X-\mu)^2\big]}{b^2}$$

Since $(X-\mu)^2\geq b^2$ if and only if $|X-\mu|\geq b$, the preceding inequality is equivalent to

$$P\big[\,|\,X-\mu\,|\geq b\,\big]\leq \frac{E\big[(X-\mu)^2\big]}{b^2}=\frac{\sigma^2}{b^2}$$

which completes the proof.

1.7.3 Laws of Large Numbers

There are two laws of large numbers that deal with the limiting behavior of random sequences. One is called the "weak" law of large numbers, and the other is called the "strong" law of large numbers. We will discuss only the weak law of large numbers.

Proposition 1.3: Let X_1, X_2, ..., X_n be a sequence of mutually independent and identically distributed random variables, and let their mean be $E[X_k]=\mu<\infty$. Similarly, let their variance be $\sigma_{X_k}^2=\sigma^2<\infty$. Let S_n denote the sum of the n random variables; that is,

$$S_n = X_1 + X_2 + \cdots + X_n$$

Then the weak law of large numbers states that for any $\varepsilon>0$,

$$\lim_{n\to\infty} P\left[\left|\frac{S_n}{n}-\mu\right|\geq\varepsilon\right]\to 0 \tag{1.46}$$

Equivalently,

$$\lim_{n\to\infty} P\left[\left|\frac{S_n}{n}-\mu\right|<\varepsilon\right]\to 1 \tag{1.47}$$

Proof: Since $X_1, X_2, \ldots, X_n$ are independent and have the same distribution, we have that

$$\text{Var}(S_n) = n\sigma^2$$

$$\text{Var}\left(\frac{S_n}{n}\right) = \frac{n\sigma^2}{n^2} = \frac{\sigma^2}{n}$$

$$E\left[\frac{S_n}{n}\right] = \frac{n\mu}{n} = \mu$$

From Chebyshev inequality, for $\varepsilon > 0$, we have that

$$P\left[\left|\frac{S_n}{n} - \mu\right| \geq \varepsilon\right] \leq \frac{\sigma^2}{n\varepsilon^2}$$

Thus, for a fixed ε,

$$P\left[\left|\frac{S_n}{n} - \mu\right| \geq \varepsilon\right] \rightarrow 0$$

as $n \rightarrow \infty$, which completes the proof.

1.7.4   The Central Limit Theorem

The central limit theorem provides an approximation to the behavior of sums of random variables. The theorem states that as the number of independent and identically distributed random variables with finite mean and finite variance increases, the distribution of their sum becomes increasingly normal regardless of the form of the distribution of the random variables. More formally, let $X_1, X_2, \ldots, X_n$ be a sequence of mutually independent and identically distributed random variables, each of which has a finite mean μ_X and a finite variance σ_X^2. Let S_n be defined as follows:

$$S_n = X_1 + X_2 + \cdots + X_n.$$

Now, $E[S_n] = n\mu_X$ and $\sigma_{S_n}^2 = n\sigma_X^2$. Converting S_n to standard normal random variable (i.e., zero mean and variance $= 1$) we obtain

$$Y_n = \frac{S_n - E[S_n]}{\sigma_{S_n}} = \frac{S_n - n\mu_X}{\sqrt{n\sigma_X^2}} = \frac{S_n - n\mu_X}{\sigma_X \sqrt{n}} \tag{1.48}$$

The central limit theorem states that if $F_{Y_n}(y)$ is the CDF of Y_n, then

$$\lim_{n \to \infty} F_{Y_n}(y) = \lim_{n \to \infty} P[Y_n \leq y] = \frac{1}{\sqrt{2\pi}} \int_{-\infty}^{y} e^{-u^2/2} \, du = \Phi(y) \tag{1.49}$$

This means that $\lim_{n\to\infty} Y_n \sim N(0,1)$. Thus, one of the important roles that the normal distribution plays in statistics is its usefulness as an approximation of other probability distribution functions.

An alternate statement of the theorem is that in the limit as n becomes very large,

$$Z_n = \frac{X_1 + X_2 + \cdots + X_n}{\sigma_X \sqrt{n}}$$

is a normal random variable with unit variance.

PROBLEMS

1.1 A sequence of Bernoulli trials consists of choosing components at random from a batch of components. A selected component is classified as either defective or nondefective. A nondefective component is considered to be a success, while a defective component is considered to be a failure. If the probability that a selected component is nondefective is 0.8, determine the probabilities of the following events:

1. The first success occurs on the fifth trial.
2. The third success occurs on the eighth trial.
3. There are 2 successes by the fourth trial, there are 4 successes by the tenth trial, and there are 10 successes by the eighteenth trial.

1.2 A lady invites 12 people for dinner at her house. Unfortunately, the dining table can only seat six people. Her plan is that if six or fewer guests come, then they will be seated at the table (i.e., they will have a sit-down dinner); otherwise, she will set up a buffet-style meal. The probability that each invited guest will come to dinner is 0.4, and each guest's decision is independent of other guests' decisions. Determine the following:

1. The probability that she has a sit-down dinner
2. The probability that she has a buffet-style dinner
3. The probability that there are at most three guests

1.3 A Girl Scout troop sells cookies from house to house. One of the parents of the girls figured out that the probability that they sell a set of packs of cookies at any house they visit is 0.4, where it is assumed that they sell exactly one set to each house that buys their cookies.

1. What is the probability that the house where they make their first sale is the fifth house they visit?
2. Given that they visited 10 houses on a particular day, what is the probability that they sold exactly six sets of cookie packs?
3. What is the probability that on a particular day the third set of cookie packs is sold at the seventh house that the girls visit?

1.4 Students arrive for a lab experiment according to a Poisson process with a rate of 12 students per hour. However, the lab attendant opens the door to the lab when at least four students are waiting at the door. What is the probability that the waiting time of the first student to arrive exceeds 20 min? (By waiting time we mean the time that elapses from when a student arrives until the door is opened by the lab attendant.)

1.5 Customers arrive at the neighborhood bookstore according to a Poisson process with an average rate of 10 customers per hour. Independently of other customers, each arriving customer buys a book with probability 1/8.

1. What is the probability that the bookstore sells no book during a particular hour?
2. What is the PDF of the time until the first book is sold?

1.6 Students arrive at the professor's office for extra help according to a Poisson process with an average rate of four students per hour. The professor does not start the tutorial until at least three students are available. Students who arrive while the tutorial is going on will have to wait for the next session.

1. Given that a tutorial has just ended and there are no students currently waiting for the professor, what is the mean time until another tutorial can start?
2. Given that one student was waiting when the tutorial ended, what is the probability that the next tutorial does not start within the first 2 h?

1.7 Consider 100 bulbs whose light times are independent and identically distributed exponential random variables with a mean of 5 h. Assume that these bulbs are used one at time and a failed bulb is immediately replaced by a new one. What is the probability that there is still a working bulb after 525 h?

1.8 Assume that X and Y are independent and identically distributed standard normal random variables. Compute the PDF of $V = 1 + X + Y$.

1.9 Assume that we roll a fair die one thousand times. Assume that the outcomes of the rolls are independent and let X be the number of fours that appear. Find the probability that $160 \leq X \leq 190$.

CHAPTER 2

OVERVIEW OF STOCHASTIC PROCESSES

2.1 INTRODUCTION

Stochastic processes deal with the dynamics of probability theory. The concept of stochastic processes enlarges the random variable concept to include time. Thus, instead of thinking of a random variable X that maps an event $w \in \Omega$, where Ω is the sample space, to some number $X(w)$, we think of how the random variable maps the event to different numbers at different times. This implies that instead of the number $X(w)$, we deal with $X(t, w)$, where $t \in T$ and T is called the *parameter set* of the process and is usually a set of times.

Stochastic processes are widely encountered in such fields as communications, control, management science, and time series analysis. Examples of stochastic processes include the population growth, the failure of an equipment, the price of a given stock over time, and the number of calls that arrive at a switchboard.

If we fix the sample point w, $X(t)$ is some real function of time. For each w, we have a function $X(t)$. Thus, $X(t, w)$ can be viewed as a collection of time functions, one for each sample point w. On the other hand, if we fix t, we have a function $X(w)$ that depends only on w and thus is a random variable. Thus, a stochastic process becomes a random variable when time is fixed at some particular value. With many values of t we obtain a collection of random variables. Thus, we can define a stochastic process as a family of random variables $\{X(t, w) | t \in T, w \in \Omega\}$ defined over a given probability space and indexed by the time parameter t. Alternatively we can define a stochastic

Elements of Random Walk and Diffusion Processes, First Edition. Oliver C. Ibe.
© 2013 John Wiley & Sons, Inc. Published 2013 by John Wiley & Sons, Inc.

process as a process that evolves randomly over time. A stochastic process is also called a *random process*. Thus, we will use the terms "stochastic process" and "random process" interchangeably.

2.2 CLASSIFICATION OF STOCHASTIC PROCESSES

A stochastic process can be classified according to the nature of the time parameter and the values that $X(t, w)$ can take on. As discussed earlier, T is called the parameter set of the random process. If T is an interval of real numbers and hence is continuous, the process is called a *continuous-time* stochastic process. Similarly, if T is a countable set and hence is discrete, the process is called a *discrete-time* random process. A discrete-time stochastic process is also called a *random sequence*, which is denoted by $\{X[n] | n = 1, 2, \ldots\}$.

The values that $X(t, w)$ assumes are called the *states* of the stochastic process. The set of all possible values of $X(t, w)$ forms the *state space*, Ω, of the stochastic process. If Ω is continuous, the process is called a *continuous-state* stochastic process. Similarly, if Ω is discrete, the process is called a *discrete-state* stochastic process. In the remainder of this book, we suppress the space argument w and denote a stochastic process by $X(t)$.

2.3 MEAN AND AUTOCORRELATION FUNCTION

The mean of the stochastic process $X(t)$ is a function of time called the ensemble average and denoted by

$$\mu(t) = E[X(t)] \tag{2.1}$$

The autocorrelation function provides a measure of the similarity between two observations of a stochastic process $\{X(t)\}$ at different points in time t and s. The autocorrelation function of $X(t)$ and $X(s)$ is denoted by $R_{xx}(t, s)$ and is defined as follows:

$$R_{XX}(t, s) = E[X(t)X(s)] = E[X(s)X(t)] = R_{XX}(s, t) \tag{2.2a}$$

$$R_{XX}(t, t) = E[X^2(t)] \tag{2.2b}$$

It is common to define $s = t + \tau$, which gives the autocorrelation function as

$$R_{XX}(t, t + \tau) = E[X(t)X(t + \tau)]$$

The parameter τ is sometimes called the *delay time* (or *lag time*). The autocorrelation function of a deterministic periodic function with period T is given by

$$R_{XX}(t, t + \tau) = \frac{1}{2T} \int_{-T}^{T} f_X(t) f_X(t + \tau) dt$$

Similarly, for an aperiodic function, the autocorrelation function is given by

$$R_{XX}(t, t + \tau) = \int_{-\infty}^{\infty} f_X(t) f_X(t + \tau) dt$$

The autocorrelation function basically defines how much a signal is similar to a time-shifted version of itself. A stochastic process $\{X(t)\}$ is called a second-order process if $R_{XX}(t, t) = E[X^2(t)] < \infty$ for each $t \in T$.

2.4 STATIONARY PROCESSES

There are several ways to define a stationary stochastic process. At a high level, it is a process whose statistical properties do not vary with time. The two widely studied types of stationary processes are the *strict-sense stationary* processes and the *wide-sense stationary* (WSS) processes.

2.4.1 Strict-Sense Stationary Processes

A stochastic process is defined to be a strict-sense stationary process if its cumulative distribution function (CDF) is invariant to a shift in the time origin. This means that the process $\{X(t)\}$ with the CDF $F_{X(t)}(x_1, x_2, \ldots, x_n; t_1, t_2, \ldots, t_n)$ is a strict-sense stationary process if its CDF is identical to that of $\{X(t + \varepsilon)\}$ for any arbitrary ε. Thus, we have that being a strict-sense stationary process implies that for any arbitrary ε,

$$F_{X(t)}(x_1, x_2, \ldots, x_n; t_1, t_2, \ldots, t_n) = F_{X(t)}(x_1, x_2, \ldots, x_n; t_1 + \varepsilon, t_2 + \varepsilon, \ldots, t_n + \varepsilon)$$

for all n. When the CDF is differentiable, the equivalent condition for strict-sense stationarity is that the probability density function (PDF) is invariant to a shift in the time origin; that is,

$$f_{X(t)}(x_1, x_2, \ldots, x_n; t_1, t_2, \ldots, t_n) = f_{X(t)}(x_1, x_2, \ldots, x_n; t_1 + \varepsilon, t_2 + \varepsilon, \ldots, t_n + \varepsilon)$$

for all n. If $\{X(t)\}$ is a strict-sense stationary process, then the CDF $F_{X_1 X_2}(x_1, x_2; t_1, t_1 + \tau)$ does not depend on t but it may depend on τ. Thus, if $t_2 = t_1 + \tau$, then it may depend on $t_2 - t_1$, but not on t_1 and t_2 individually. This means that if $\{X(t)\}$ is a strict-sense stationary process, then the autocorrelation and autocovariance functions do not depend on t. Thus, we have that for all $\tau \in T$

$$\mu_X(t) = \mu_X(0) \tag{2.3a}$$

$$R_{XX}(t, t + \tau) = R_{XX}(0, \tau) \tag{2.3b}$$

$$C_{XX}(t, t + \tau) = C_{XX}(0, \tau) \tag{2.3c}$$

where $C_{XX}(t_1, t_2) = E[\{X(t_1) - \mu_X(t_1)\}\{X(t_2) - \mu_X(t_2)\}]$ is the autocovariance function. If the condition $\mu_X(t) = \mu_X$ holds for all t, the mean is constant and denoted by μ_X. Similarly, if the function $R_{XX}(t, t + \tau)$ does not depend on t but is a function of τ, we write $R_{XX}(0, \tau) = R_{XX}(\tau)$. Finally, whenever the condition $C_{XX}(t, t + \tau) = C_{XX}(0, \tau)$ holds for all t, we write $C_{XX}(0, \tau) = C_{XX}(\tau)$.

2.4.2 Wide-Sense Stationary Processes

Many practical problems that we encounter require that we deal with only the mean and autocorrelation function of a random process. Solutions to these problems are simplified if these quantities do not depend on absolute time. Stochastic processes in which the mean and autocorrelation function do not depend on absolute time are called WSS processes. Thus, for a WSS process $\{X(t)\}$,

$$E[X(t)] = \mu_X \tag{2.4a}$$

$$R_{XX}(t, t + \tau) = R_{XX}(\tau) \tag{2.4b}$$

Note that a strict-sense stationary process is also a WSS process. However, in general the converse is not true; that is, a WSS process is not necessarily stationary in the strict sense.

2.5 POWER SPECTRAL DENSITY

For a WSS stochastic process $\{X(t)\}$, the Fourier transform of the autocorrelation function is called the *power spectral density*, $S_{XX}(w)$, of the process. Thus,

$$S_{XX}(w) = \int_{-\infty}^{\infty} e^{-jw\tau} R_{XX}(\tau) d\tau \tag{2.5}$$

We can recover $R_{XX}(\tau)$ via the inverse Fourier transform operation as follows:

$$R_{XX}(\tau) = \frac{1}{2\pi} \int_{-\infty}^{\infty} S_{XX}(w) e^{jw\tau} dw \tag{2.6}$$

The mean-square value $E[X^2(t)]$ of the stochastic process $\{X(t)\}$ is related to the power spectral density as follows:

$$E[X^2(t)] = R_{XX}(0) = \frac{1}{2\pi} \int_{-\infty}^{\infty} S_{XX}(w) dw \tag{2.7}$$

Other properties of $S_{XX}(w)$ include the fact that it is a nonnegative function; that is, $S_{XX}(w) \geq 0$. Also, it is an even function; that is, $S_{XX}(-w) = S_{XX}(w)$.

2.6   COUNTING PROCESSES

A stochastic process $\{X(t)\}$ is called a counting process if $X(t)$ represents the total number of "events" that have occurred in the interval $[0, t]$. An example of a counting process is the number of customers that arrive at a bank from the time the bank opens its doors for business until some time t. A counting process satisfies the following conditions:

1. $X(t) \geq 0$, which means that it has nonnegative values.
2. $X(0) = 0$, which means that the counting of events begins at time 0.
3. $X(t)$ is integer-valued.
4. If $s < t$, then $X(s) \leq X(t)$, which means that it is a nondecreasing function of time.
5. $X(t) - X(s)$ represents the number of events that have occurred in the interval $[s, t]$.

2.7   INDEPENDENT INCREMENT PROCESSES

A counting process is defined to be an independent increment process if the number of events that occur in disjoint time intervals is an independent random variable. Thus, $\{X(t)\}$ is an independent increment process if for every set of time instants $t_0 = 0 < t_1 < t_2 < \cdots < t_n$ the increments $X(t_1) - X(t_0)$, $X(t_2) - X(t_1)$, ..., $X(t_n) - X(t_{n-1})$ are mutually independent random variables.

2.8   STATIONARY INCREMENT PROCESS

A counting process $\{X(t)\}$ is defined to possess stationary increments if for every set of time instants $t_0 = 0 < t_1 < t_2 < \ldots < t_n$ the increments $X(t_1) - X(t_0)$, $X(t_2) - X(t_1)$, ..., $X(t_n) - X(t_{n-1})$ are identically distributed. In general, the mean of an independent increment process $\{X(t)\}$ with stationary increments has the form

$$E[X(t)] = mt \tag{2.8}$$

where the constant m is the value of the mean at time $t = 1$. That is, $m = E[X(1)]$. Similarly, the variance of an independent increment process $\{X(t)\}$ with stationary increments has the form

$$\text{Var}[X(t)] = \sigma^2 t \tag{2.9}$$

where the constant σ^2 is the value of the variance at time $t = 1$; that is, $\sigma^2 = \text{Var}[X(1)]$.

2.9 POISSON PROCESSES

Poisson processes are widely used to model arrivals (or occurrence of events) in a system. For example, they are used to model the arrival of telephone calls at a switchboard, the arrival of customers' orders at a service facility, and the random failures of equipment. There are two ways to define a Poisson process. The first definition of the process is that it is a counting process $\{X(t), t \geq 0\}$ in which the number of events in any interval of length t has a Poisson distribution with mean λt. Thus, for all $s, t > 0$,

$$P[X(s+t) - X(s) = n] = \frac{(\lambda t)^n}{n!} e^{-\lambda t} \quad n = 0, 1, 2, \ldots$$

The second way to define the Poisson process $X(t)$ is that it is a counting process with stationary and independent increments such that for a rate $\lambda > 0$ the following conditions hold:

1. $P[X(t+\Delta t) - X(t) = 1] = \lambda \Delta t + o(\Delta t)$, which means that the probability of one event within a small time interval Δt is approximately $\lambda \Delta t$, where $o(\Delta t)$ is a function of Δt that goes to zero faster than Δt does. That is,

$$\lim_{\Delta t \to 0} \frac{o(\Delta t)}{\Delta t} = 0$$

2. $P[X(t+\Delta t) - X(t) \geq 2] = o(\Delta t)$, which means that the probability of two or more events within a small time interval Δt is $o(\Delta t)$.

3. $P[X(t+\Delta t) - X(t) = 0] = 1 - \lambda \Delta t + o(\Delta t)$

These three properties enable us to derive the probability mass function (PMF) of the number of events in a time interval of length t as follows:

$$P[X(t+\Delta t) = n] = P[X(t) = n]P[X(\Delta t) = 0]$$
$$+ P[X(t) = n-1]P[X(\Delta t) = 1] + o(\Delta t)$$
$$= P[X(t) = n](1 - \lambda \Delta t) + P[X(t) = n-1]\lambda \Delta t + o(\Delta t)$$
$$P[X(t+\Delta t) = n] - P[X(t) = n] = -\lambda P[X(t) = n]\Delta t + \lambda P[X(t) = n-1]\Delta t + o(\Delta t)$$
$$\frac{P[X(t+\Delta t) = n] - P[X(t) = n]}{\Delta t} = -\lambda P[X(t) = n] + \lambda P[X(t) = n-1] + \frac{o(\Delta t)}{\Delta t}$$
$$\lim_{\Delta t \to 0} \left\{ \frac{P[X(t+\Delta t) = n] - P[X(t) = n]}{\Delta t} \right\} = \frac{d}{dt} P[X(t) = n]$$
$$= -\lambda P[X(t) = n] + \lambda P[X(t) = n-1]$$

That is,

$$\frac{d}{dt}P[X(t)=n]+\lambda P[X(t)=n]=\lambda P[X(t)=n-1] \tag{2.10}$$

This equation may be solved iteratively for $n=0, 1, 2, \ldots$, subject to the initial conditions

$$P[X(0)=n]=\begin{cases}1 & n=0 \\ 0 & n\neq 0\end{cases}$$

This gives the PMF of the number of events (or "arrivals") in an interval of length t as

$$p_{X(t)}(n,t)=\frac{(\lambda t)^n}{n!}e^{-\lambda t} \quad t\geq 0; \; n=0,1,2,\ldots$$

From the results obtained for Poisson random variables earlier in Chapter 1, we have that

$$G_{X(t)}(z)=e^{-\lambda t(1-z)}$$
$$E[X(t)]=\lambda t$$
$$\sigma_{X(t)}^2=\lambda t$$

The fact that the mean is $E[X(t)]=\lambda t$ indicates that λ is the expected number of arrivals per unit time in the Poisson process. Thus, the parameter λ is called the *arrival rate* for the process. If λ is independent of time, the Poisson process is called a *homogeneous Poisson process*. Sometimes the arrival rate is a function of time, and we represent it as $\lambda(t)$. Such processes are called *nonhomogeneous Poisson processes*.

Another important property of the Poisson process is that the interarrival times of customers are exponentially distributed with parameter λ. This can be demonstrated by noting that if T is the time until the next arrival, then $P[T>t]=P[X(t)=0]=e^{-\lambda t}$ since there is no arrival by time t. Therefore, the CDF and PDF of the interarrival time are given by

$$F_T(t)=P[T\leq t]=1-P[T>t]=1-e^{-\lambda t}$$
$$f_T(t)=\frac{d}{dt}F_T(t)=\lambda e^{-\lambda t}$$

Thus, the time until the next arrival is always exponentially distributed. From the memoryless property of the exponential distribution, the future evolution of the Poisson process is independent of the past and is always probabilistically the same. Therefore, the Poisson process is memoryless. As the saying goes, "the Poisson process implies exponential distribution, and the exponential distribution implies the Poisson process." The Poisson process deals with the number of arrivals within a given time interval where interarrival times are exponentially distributed, and the exponential distribution measures the time between arrivals where customers arrive according to a Poisson process.

2.9.1   Compound Poisson Process

Let $\{N(t)|t \geq 0\}$ be a Poisson process with arrival rate λ. Let $\{Y_i, i = 1, 2, \ldots\}$ be a family of independent and identically distributed random variables. Assume that the Poisson process $\{N(t)|t \geq 0\}$ and the sequence $\{Y_i, i = 1, 2, \ldots\}$ are independent. We define a stochastic process $\{X(t)|t \geq 0\}$ to be a *compound Poisson process* if, for $t \geq 0$, it can be represented by

$$X(t) = \sum_{i=1}^{N(t)} Y_i \qquad (2.11)$$

Thus, $X(t)$ is a Poisson sum of independent and identically distributed random variables. One example of the concept of compound Poisson process is the following. Assume students arrive at the university bookstore to buy books in a Poisson manner. If the number of books that each of these students buys is an independent and identically distributed random variable, then the number of books bought by time t is a compound Poisson process.

Another way to look at the compound Poisson process is that it is a Poisson process in which the jump sizes, instead of being of size 1, are allowed to be independent and identically distributed with a general distribution. It is useful in modeling queues and the insurance business. In the case of insurance, $X(t)$ can denote the total claim damages that arrive by time t; and in the case of queues, it can denote the total amount of work that arrived to the queueing system by time t.

Because the compound Poisson process has a rate that takes on a stochastic nature, it is also called a *doubly stochastic Poisson process*. This term is used to emphasize the fact that the process involves two kinds of randomness: There is a randomness that is associated with the main process that is sometimes called the *Poisson point process*, and there is another independent randomness that is associated with its rate.

Assume that the Y_i are discrete random variables with the PMF $p_Y(y)$. The value of $X(t)$, given that $N(t) = n$, is $X(t)|_{N(t)=n} = Y_1 + Y_2 + \ldots + Y_n$. Thus, the conditional z-transform of the PMF of $X(t)$, given that $N(t) = n$, is given by

$$G_{X(t)|N(t)}(z \mid n) = E\left[z^{Y_1 + Y_2 + \cdots + Y_n}\right] = \left(E\left[z^Y\right]\right)^n = \left[G_Y(z)\right]^n$$

where the last two equalities follow from the fact that the Y_i are independent and identically distributed. Thus, the unconditional z-transform of the PMF of $X(t)$ is given by

$$G_{X(t)}(z) = \sum_{n=0}^{\infty} G_{X(t)|N(t)}(z \mid n)\, p_{N(t)}(n) = \sum_{n=0}^{\infty} \left[G_Y(z)\right]^n p_{N(t)}(n)$$

$$= \sum_{n=0}^{\infty} \left[G_Y(z)\right]^n \frac{(\lambda t)^n}{n!} e^{-\lambda t} = e^{-\lambda t} \sum_{n=0}^{\infty} \frac{\left[\lambda t G_Y(z)\right]^n}{n!} = e^{-\lambda t} e^{\lambda t G_Y(z)}$$

$$= e^{-\lambda t [1 - G_Y(z)]} \qquad (2.12)$$

The mean and variance of $X(t)$ can be obtained through differentiating the aforementioned function. These are given by

$$E[X(t)] = \frac{d}{dz}G_{X(t)}(z)\mid_{z=1} = \lambda t E[Y] \qquad (2.13a)$$

$$E[X^2(t)] = \frac{d^2}{dz^2}G_{X(t)}(z)\mid_{z=1} + \frac{d}{dz}G_{X(t)}(z)\mid_{z=1} = \lambda t E[Y^2] + \{\lambda t E[Y]\}^2$$

$$\sigma^2_{X(t)} = E[X^2(t)] - \{E[X(t)]\}^2 = \lambda t E[Y^2] \qquad (2.13b)$$

If the Y_i are continuous random variables, we would obtain the s-transform of the PDF of $X(t)$ as $M_{X(t)}(s) = G_N(M_Y(s)) = e^{-\lambda t[1-M_Y(s)]}$ and the earlier results still hold.

2.10 MARKOV PROCESSES

Markov processes are widely used in engineering, physical science, social science, and business systems modeling. They are used to model systems that have a limited memory of their past. For example, consider a sequence of games where a player gets \$1 if he wins a game and loses \$1 if he loses the game. Then the amount of money the player will make after $n+1$ games is determined by the amount of money he has made after n games. Any other information is irrelevant in making this prediction. In population growth studies, the population of the next generation depends mainly on the current population and possibly the last few generations.

A random process $\{X(t)|t \in T\}$ is called a first-order Markov process if for any $t_0 < t_1 < t_2 < \cdots < t_n$ the conditional CDF of $X(t_n)$ for given values of $X(t_0)$, $X(t_1)$, ..., $X(t_{n-1})$ depends only on $X(t_{n-1})$. That is,

$$P[X(t_n) \leq x_n \mid X(t_{n-1}) = x_{n-1}, X(t_{n-2}) = x_{n-2}, \ldots, X(t_0) = x_0]$$
$$= P[X(t_n) \leq x_n \mid X(t_{n-1}) = x_{n-1}] \qquad (2.14)$$

This means that, given the present state of the process, the future state is independent of the past. This property is usually referred to as the *Markov property*. In second-order Markov processes, the future state depends on both the current state and the last immediate state, and so on for higher-order Markov processes. In this chapter, we consider only first-order Markov processes.

Markov processes are classified according to the nature of the time parameter and the nature of the state space. With respect to state space, a Markov process can be either a discrete-state Markov process or continuous-state Markov process. A discrete-state Markov process is called a *Markov chain*. Similarly, with respect to time, a Markov process can be either a discrete-time Markov process or a continuous-time Markov process. Thus, there are four basic types of Markov processes:

- Discrete-time Markov chain (or discrete-time discrete-state Markov process)
- Continuous-time Markov chain (or continuous-time discrete-state Markov process)
- Discrete-time Markov process (or discrete-time continuous-state Markov process)
- Continuous-time Markov process (or continuous-time continuous-state Markov process)

2.10.1   Discrete-Time Markov Chains

The discrete-time process $\{X_k | k = 0, 1, 2, \ldots\}$ is called a Markov chain if for all $i, j, k, \ldots, m$, the following is true:

$$P[X_k = j \mid X_{k-1} = i, \; X_{k-2} = \alpha, \ldots, X_0 = \theta] = P[X_k = j \mid X_{k-1} = i] = p_{ijk} \quad (2.15)$$

The quantity p_{ijk} is called the *state transition probability*, which is the conditional probability that the process will be in state j at time k immediately after the next transition, given that it is in state i at time $k-1$. A Markov chain that obeys the preceding rule is called a *nonhomogeneous Markov chain*. In this book, we will consider only *homogeneous Markov chains*, which are Markov chains in which $p_{ijk} = p_{ij}$. This means that homogeneous Markov chains do not depend on the time unit, which implies that

$$P[X_k = j \mid X_{k-1} = i, \; X_{k-2} = \alpha, \ldots, X_0 = \theta] = P[X_k = j \mid X_{k-1} = i] = p_{ij} \quad (2.16)$$

This is the so-called *Markov property*. The *homogeneous state transition probability* satisfies the following conditions:

- $0 \leq p_{ij} \leq 1$
- Because the states are mutually exclusive and collectively exhaustive we have that $\sum_j p_{ij} = 1$, $i = 1, 2, \ldots, n$.

From the aforementioned definition we obtain the following *Markov chain rule*:

$$\begin{aligned}
P[X_k &= j, X_{k-1} = i_1, X_{k-2}, \ldots, X_0] \\
&= P[X_k = j \mid X_{k-1} = i_1, X_{k-2} = i_2, \ldots, X_0] P[X_{k-1} = i, X_{k-2}, \ldots, X_0] \\
&= P[X_k = j \mid X_{k-1} = i_1] P[X_{k-1} = i_2, X_{k-2}, \ldots, X_0] \\
&= P[X_k = j \mid X_{k-1} = i_1] P[X_{k-1} = i_1 \mid X_{k-2} = i_2] \ldots P[X_1 = i_{k-1} \mid X_0 = i_k] P[X_0 = i_k] \\
&= p_{i_1 j} p_{i_2 i_1} p_{i_3 i_2} \cdots p_{i_k i_{k-1}} P[X_0 = i_k]
\end{aligned}$$

$$(2.17)$$

Thus, once we know the probability of being in the initial state X_0 we can evaluate the joint probability $P[X_k, X_{k-1}, ..., X_0]$.

2.10.2 State Transition Probability Matrix

It is customary to display the state transition probabilities as the entries of an $n \times n$ matrix P, where p_{ij} is the entry in the ith row and jth column:

$$P = \begin{bmatrix} p_{11} & p_{12} & \cdots & p_{1n} \\ p_{21} & p_{22} & \cdots & p_{2n} \\ \cdots & \cdots & \cdots & \cdots \\ p_{n1} & p_{n2} & \cdots & p_{nn} \end{bmatrix}$$

P is called the *transition probability matrix*. It is a *stochastic matrix* because for any row i, $\sum_j p_{ij} = 1$.

2.10.3 The k-Step State Transition Probability

Let $p_{ij}(k)$ denote the conditional probability that the system will be in state j after exactly k transitions, given that it is presently in state i. That is,

$$p_{ij}(k) = P\left[X_{m+k} = j \mid X_m = i\right]$$
$$p_{ij}(0) = \begin{cases} 1 & i = j \\ 0 & i \neq j \end{cases}$$
$$p_{ij}(1) = p_{ij}$$

Consider the two-step transition probability $p_{ij}(2)$, which is defined by

$$p_{ij}(2) = P\left[X_{m+2} = j \mid X_m = i\right]$$

Assume that $m=0$, then

$$p_{ij}(2) = P[X_2 = j \mid X_0 = i] = \sum_l P\left[X_2 = j, X_1 = l \mid X_0 = i\right]$$
$$= \sum_l P\left[X_2 = j \mid X_1 = l, X_0 = i\right] P\left[X_1 = l \mid X_0 = i\right]$$
$$= \sum_l P\left[X_2 = j \mid X_1 = l\right] P\left[X_1 = l \mid X_0 = i\right]$$
$$= \sum_l p_{lj} p_{il}$$

where the second to the last equality is due to the Markov property. The final equation states that the probability of starting in state i and being in state j at the end of the second transition is the probability that we first go immediately from state i to an intermediate state l and then immediately from state l to state j; the summation is taken over all possible intermediate states l.

The following proposition deals with a class of equations called the *Chapman-Kolmogorov equations*, which provide a generalization of the earlier results obtained for the two-step transition probability.

Proposition 2.1: For all $0<r<n$,

$$p_{ij}(k) = \sum_l p_{ij}(r)p_{il}(k-r) \tag{2.18}$$

This proposition states that the probability that the process starts in state i and finds itself in state j at the end of the kth transition is the product of the probability that the process starts in state i and finds itself in an intermediate state l after r transitions and the probability that it goes from state l to state j after additional $k-r$ transitions.

2.10.4   State Transition Diagrams

Consider the following problem. It has been observed via a series of tosses of a particular biased coin that the outcome of the next toss depends on the outcome of the current toss. In particular, given that the current toss comes up heads, the next toss will come up heads with probability 0.6 and tails with probability 0.4. Similarly, given that the current toss comes up tails, the next toss will come up heads with probability 0.35 and tails with probability 0.65.

If we define state 1 to represent heads and state 2 to represent tails, then the transition probability matrix for this problem is the following:

$$P = \begin{bmatrix} 0.6 & 0.4 \\ 0.35 & 0.65 \end{bmatrix}$$

All the properties of the Markov process can be determined from this matrix. However, the analysis of the problem can be simplified by the use of the *state-transition diagram* in which the states are represented by circles and directed arcs represent transitions between states. The state transition probabilities are labeled on the appropriate arcs. Thus, with respect to the aforementioned problem, we obtain the state-transition diagram shown in Figure 2.1.

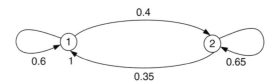

FIGURE 2.1   Example of state-transition diagram.

2.10.5 Classification of States

A state j is said to be *accessible* (or *can be reached*) from state i if, starting from state i, it is possible that the process will ever enter state j. This implies that $p_{ij}(n) > 0$ for some $n > 0$. Thus, the n-step probability enables us to obtain reachability information between any two states of the process.

Two states that are accessible from each other are said to *communicate* with each other. The concept of communication divides the state space into different classes. Two states that communicate are said to be in the same *class*. All members of one class communicate with one another. If a class is not accessible from any state outside the class, we define the class to be a *closed communicating class*. A Markov chain in which all states communicate, which means that there is only one class, is called an *irreducible* Markov chain. For example, the Markov chain shown in Figure 2.1 is an irreducible chain.

The states of a Markov chain can be classified into two broad groups: those that the process enters infinitely often and those that it enters finitely often. In the long run, the process will be found to be in only those states that it enters infinitely often. Let $f_{ij}(n)$ denote the conditional probability that given that the process is presently in state i, the first time it will enter state j occurs in exactly n transitions (or steps). We call $f_{ij}(n)$ the probability of *first passage* from state i to state j in n transitions. The parameter f_{ij}, which is defined as follows:

$$f_{ij} = \sum_{n=1}^{\infty} f_{ij}(n) \tag{2.19}$$

is the probability of first passage from state i to state j. It is the conditional probability that the process will ever enter state j, given that it was initially in state i. Obviously, $f_{ij}(1) = p_{ij}$ and a recursive method of computing $f_{ij}(n)$ is

$$f_{ij}(n) = \sum_{l \neq j} p_{il} f_{lj}(n-1) \tag{2.20}$$

The quantity f_{ii} denotes the probability that a process that starts at state i will ever return to state i. Any state i for which $f_{ii} = 1$ is called a *recurrent state*, and any state i for which $f_{ii} < 1$ is called a *transient state*. More formally, we define these states as follows:

- A state j is called a *transient* (or *nonrecurrent*) state if there is a positive probability that the process will never return to j again after it leaves j.
- A state j is called a *recurrent* (or *persistent*) state if, with probability 1, the process will eventually return to j after it leaves j. A set of recurrent states forms a *single chain* if every member of the set communicates with all other members of the set.

- A recurrent state j is called a *periodic* state if there exists an integer d, $d > 1$, such that $p_{jj}(n)$ is 0 for all values of n other than $d, 2d, 3d, \ldots$; d is called the period. If $d = 1$, the recurrent state j is said to be *aperiodic*.
- A recurrent state j is called a *positive recurrent* state if starting at state j the expected time until the process returns to state j is finite. Otherwise, the recurrent state is called a *null recurrent* state.
- Positive recurrent, aperiodic states are called *ergodic* states.
- A chain consisting of ergodic states is called an *ergodic chain*.
- A state j is called an *absorbing* (or *trapping*) state if $p_{jj} = 1$. Thus, once the process enters a trapping or absorbing state, it never leaves the state, which means that it is "trapped."

2.10.6 Limiting-State Probabilities

Recall that the k-step state transition probability $p_{ij}(k)$ is the conditional probability that the system will be in state j after exactly k transitions, given that it is presently in state i. The k-step transition probabilities can be obtained by multiplying the transition probability matrix by itself k times. It has been found that as we multiply the transition probability matrix by itself many times, the entries remain constant. More importantly, all the members of one column will tend to converge to the same value.

For a large class of Markov chains, it can be shown that as $k \to \infty$, the k-step transition probability $p_{ij}(k)$ does not depend on i, which means that the probability that the process is in state j after k steps, $P[X(k) = j]$, approaches a constant that is independent of the initial conditions. For any Markov chain in which this limit exists, we define the *limiting-state probabilities* as follows:

$$\lim_{k \to \infty} P[X(k) = j] = \pi_j \quad j = 1, 2, \ldots, N \tag{2.21}$$

Recall that the k-step transition probability can be written in the form

$$p_{ij}(k) = \sum_l p_{il}(k-1) p_{lj}$$

If the limiting-state probabilities exist and do not depend on the initial state, then

$$\lim_{k \to \infty} p_{ij}(k) = \pi_j = \lim_{k \to \infty} \sum_l p_{il}(k-1) p_{lj} = \sum_l \pi_l p_{lj} \tag{2.22}$$

A system of linear equations that the π_i must satisfy is as follows:

$$\pi_j = \sum_l \pi_l p_{lj} \tag{2.23a}$$

$$1 = \sum_j \pi_j \tag{2.23b}$$

The following proposition specifies the conditions for the existence of the limiting-state probabilities:

Proposition 2.2:

- In any irreducible, aperiodic Markov chain, the limits $\pi_j = \lim\limits_{n \to \infty} p_{ij}(n)$ exist and are independent of the initial distribution.
- In any irreducible, periodic Markov chain, the limits $\pi_j = \lim\limits_{n \to \infty} p_{ij}(n)$ exist and are independent of the initial distribution. However, they must be interpreted as the long-run probability that the process is in state j.

2.10.7 Doubly Stochastic Matrix

A transition probability matrix P is defined to be a doubly stochastic matrix if each of its columns sums to 1. That is, not only does each row sum to 1, each column also sums to 1. Thus, for every column j of a doubly stochastic matrix, we have that $\sum_i p_{ij} = 1$.

Doubly stochastic matrices have interesting limiting-state probabilities, as the following theorem shows.

Theorem 2.1: If P is a doubly stochastic matrix associated with the transition probabilities of a Markov chain with N states, then the limiting-state probabilities are given by $\pi_i = 1/N$, $i = 1, 2, \ldots, N$.

2.10.8 Continuous-Time Markov Chains

A stochastic process $\{X(t)|t \geq 0\}$ is a continuous-time Markov chain if, for all $s, t \geq 0$ and nonnegative integers i, j, k,

$$P\left[X(t+s) = j \mid X(s) = i, \ X(u) = k, \ 0 \leq u \leq s\right] = P\left[X(t+s) = j \mid X(s) = i\right]$$

$$(2.24)$$

This means that in a continuous-time Markov chain, the conditional probability of the future state at time $t+s$, given the present state at the current time s and all past states, depends only on the present state and is independent of the past. If, in addition $P[X(t+s)=j|X(s)=i]$ is independent of s, then the process $\{X(t)|t \geq 0\}$ is said to be *time homogeneous* or have the *time homogeneity property*. Time homogeneous Markov chains have stationary (or homogeneous) transition probabilities. Let

$$p_{ij}(t) = P\left[X(t+s) = j \mid X(s) = i\right] \tag{2.25a}$$

$$p_j(t) = P\left[X(t) = j\right] \tag{2.25b}$$

That is, $p_{ij}(t)$ is the probability that a Markov chain that is presently in state i will be in state j after an additional time t, and $p_j(t)$ is the probability that a Markov chain is in state j at time t. Thus, the $p_{ij}(t)$ are the *transition probability functions* that satisfy the following condition:

$$\sum_j p_{ij}(t) = 1 \quad 0 \le p_{ij}(t) \le 1$$

Also,

$$\sum_j p_j(t) = 1$$

which follows from the fact that at any given time the process must be in some state. It can be shown that

$$p_{ij}(t+s) = \sum_l p_{il}(t) p_{lj}(s) \tag{2.26}$$

This equation is called the Chapman–Kolmogorov equation for the continuous-time Markov chain.

Whenever a continuous-time Markov chain enters a state i, it spends an amount of time called the *dwell time* (or *holding time*) in that state. The holding time in state i is exponentially distributed with mean $1/v_i$. At the expiration of the holding time, the process makes a transition to another state j with probability p_{ij}, where $\sum_j p_{ij} = 1$. Because the mean holding time in state i is $1/v_i$, v_i represents the rate at which the process leaves state i and $v_i p_{ij}$ represents the rate when in state i that the process makes a transition to state j. If we define v_{ij} as the rate at which the process makes a transition to state j when it is state i, then $v_{ij} = v_i p_{ij} \Rightarrow p_{ij} = v_{ij}/v_i$. If we assume that in the steady state $p_j(t) \to p_j$, then it can be shown that

$$p_i \sum_{j \ne i} v_{ij} = \sum_{j \ne i} p_j v_{ji} \quad i = 1, 2, \ldots, N \tag{2.27a}$$

$$\sum_{i=1}^{N} p_i = 1 \tag{2.27b}$$

The left side of Equation 2.27a is the rate of transition out of state i, while the right hand is the rate of transition into state i. This "balance" equation states that in the steady state the two rates are equal for any state in the Markov chain.

2.10.9 Birth and Death Processes

Birth and death processes are a special type of continuous-time Markov chains. Consider a continuous-time Markov chain with states 0, 1, 2, ... If $p_{ij} = 0$ whenever $j \ne i-1$ or $j \ne i+1$, then the Markov chain is called a birth and death process. Thus, a birth and death process is a continuous-time Markov chain with states 0, 1, 2, ... in which transitions from state i can only go to either state $i-1$ or state $i+1$. That is, a transition either causes an increase in state by one or a decrease in state by one. A birth is said to occur when the state increases by one, and a death is said to occur when the state decreases by one. For a birth and death process, we define the following *transition rates* from state i:

$$\lambda_i = v_{i(i+1)} = v_i p_{i(i+1)} \tag{2.28a}$$

$$\mu_i = v_{i(i-1)} = v_i p_{i(i-1)} \tag{2.28b}$$

Thus, λ_i is the rate at which a birth occurs when the process is in state i and μ_i is the rate at which a death occurs when the process is in state i. The sum of these two rates is $\lambda_i + \mu_i = v_i$, which is the rate of transition out of state i. The *state-transition-rate diagram* of a birth and death process is shown in Figure 2.2. It is called a state-transition-rate diagram as opposed to a state-transition diagram because it shows the rate at which the process moves from state to state and not the probability of moving from one state to another. Note that $\mu_0 = 0$, since there can be no death when the process is in empty state.

The actual state-transition probabilities when the process is in state i are $p_{i(i+1)}$ and $p_{i(i-1)}$. By definition, $p_{i(i+1)} = \lambda_i/(\lambda_i + \mu_i)$ is the probability that a birth occurs before a death when the process is in state i. Similarly, $p_{i(i-1)} = \mu_i/(\lambda_i + \mu_i)$ is the probability that a death occurs before a birth when the process is in state i.

If we assume that the limiting probabilities exist, then from Equations 2.27a and 2.27b we obtain the following:

$$\begin{gathered} \lambda_0 p_0 = \mu_1 p_1 \\ \left(\lambda_i + \mu_i\right) p_i = \lambda_{i-1} p_{i-1} + \mu_{i+1} p_{i+1} \quad i = 1, 2, \ldots \\ \sum_i p_i = 1 \end{gathered} \tag{2.29}$$

The equation states that the rate at which the process leaves state i either through a birth or a death is equal to the rate at which it enters the state through a birth when the process is in state $i-1$ or through a death when the process is in state $i+1$. This

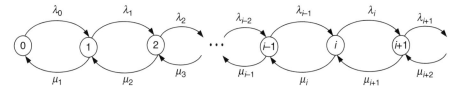

FIGURE 2.2   State-transition-rate diagram for birth and death process.

is called the *balance equation* because it balances (or equates) the rate at which the process enters state i with the rate at which it leaves state i.

2.11 GAUSSIAN PROCESSES

Gaussian processes are important in many ways. First, many physical problems are the results of adding large numbers of independent random variables. According to the central limit theorem, such sums of large numbers of random variables are essentially normal (or Gaussian) random variables. Also, the analysis of many systems is simplified if they are assumed to be Gaussian processes because of the properties of Gaussian processes. For example, noise in communication systems is usually modeled as a Gaussian process. Similarly, noise voltages in resistors are modeled as Gaussian processes.

A stochastic process $\{X(t), t \in T\}$ is defined to be a Gaussian process if and only if for any choice of n real coefficients $a_1, a_2, \ldots, a_n$ and choice of n time instants t_1, $t_2, \ldots, t_n$ in the index set T, the random variable $a_1 X(t_1) + a_2 X(t_2) + \cdots + a_n X(t_n)$ is a Gaussian (or normal) random variable. That is, $\{X(t), t \in T\}$ is a Gaussian process if any finite linear combination of $X(t)$ is a normally distributed random variable. This definition implies that the random variables $X(t_1), X(t_2), \ldots, X(t_n)$ have a jointly normally distributed PDF.

We list three important properties of Gaussian processes:

1. A Gaussian process that is WSS is also strict-sense stationary.
2. If the input to a linear system is a Gaussian process, then the output is also a Gaussian process.
3. If the input $X(t)$ to a linear system is a zero-mean Gaussian process, the output process $Y(t)$ is also a zero-mean process.

2.12 MARTINGALES

A stochastic process $\{X_n, n = 1, 2, \ldots\}$ is defined to be a martingale process (or a martingale) if it has the following properties:

- $E[|X_n|] < \infty$ for all n; that is, it has finite means, and
- $E[X_{n+1}|X_1, X_2, \ldots, X_n] = X_n$; that is, the best prediction of its future values is its present value.

If $E[X_{n+1}|X_1, X_2, \ldots, X_n] \le X_n$, then $\{X_n, n = 1, 2, \ldots\}$ is called a *supermartingale*. Similarly, if $E[X_{n+1}|X_1, X_2, \ldots, X_n] \ge X_n$, then $\{X_n, n = 1, 2, \ldots\}$ is called a *submartingale*. Thus, a martingale satisfies the conditions for both a supermartingale and a submartingale.

Sometimes the martingale property is defined with respect to another stochastic process. Specifically, let $\{X_n, n=1, 2, ...\}$ and $\{Y_n, n=1, 2, ...\}$ be stochastic processes. $\{X_n\}$ is defined to be a martingale with respect to $\{Y_n\}$ if, for $n=1, 2, ...$, the following conditions hold:

$$E[|X_n|] < \infty$$

$$E[X_{n+1} \mid Y_1, Y_2, ..., Y_n] = X_n$$

A martingale captures the essence of a fair game in the sense that regardless of a player's current and past fortunes, his expected fortune at any time in the future is the same as his current fortune. Thus, on the average he neither wins nor loses any money. Also, martingales fundamentally deal with conditional expectation. If we define $\mathfrak{I}_n = \{Y_1, Y_2, ..., Y_n\}$, then $\mathfrak{I}_n$ can be considered the potential information that is being revealed as time progresses. Therefore, we may consider a martingale as a process whose expected value, conditional on some potential information, is equal to the value revealed by the last available information.

Martingales occur in many stochastic processes. They have also become an important tool in modern financial mathematics because martingales provide one idea of fair value in financial markets.

Theorem 2.2: If $\{X_n, n \geq 0\}$ is a martingale, then $E[X_n] = E[X_0]$ for all $n \geq 0$.

Proof: Let $\mathfrak{I}_n$ be as defined earlier. We know that $E[E[X|Y]] = E[X]$. Also, because $\{X_n, n \geq 0\}$ is a martingale, $E[X_n \mid \mathfrak{I}_0] = X_0$. Thus, we have that

$$E[X_n] = E\big[E[X_n \mid \mathfrak{I}_0]\big] = E[X_0]$$

which completes the proof.

Example 2.1: Let $X_1, X_2, ...$ be independent random variables with mean 0 and let $Y_n = \sum_{k=1}^{n} X_k$. We show that the process $\{Y_n, n \geq 1\}$ is a martingale as follows:

$$
\begin{aligned}
E[Y_{n+1} \mid Y_1, ..., Y_n] &= E[Y_n + X_{n+1} \mid Y_1, ..., Y_n] \\
&= E[Y_n \mid Y_1, ..., Y_n] + E[X_{n+1} \mid Y_1, ..., Y_n] \\
&= Y_n + E[X_{n+1}] \\
&= Y_n
\end{aligned}
$$

Example 2.2: Let X_1, X_2, ... be independent random variables with mean $E[X_k] = 1$ for $k \geq 1$ and let $Y_n = \prod_{k=1}^{n} X_k$. We show that the process $\{Y_n, n \geq 1\}$ is a martingale as follows:

$$E[Y_{n+1} \mid Y_1, Y_2, ..., Y_n] = E[Y_n X_{n+1} \mid Y_1, Y_2, ..., Y_n] = Y_n E[X_{n+1} \mid Y_1, Y_2, ..., Y_n]$$
$$= Y_n E[X_{n+1}]$$
$$= Y_n$$

2.12.1 Stopping Times

Consider a stochastic process $\{X_n, n \geq 0\}$. The nonnegative integer-valued random variable T is called a stopping time for X if the event $\{T = n\}$ depends only on $\{X_0, X_1, ..., X_n\}$, and does not depend on $\{X_{n+k}, k \geq 1\}$. If T_k is a stopping time, then we have that

$$\{T_k = n\} = \{X_1 \neq k, X_2 \neq k, ..., X_{n-1} \neq k, X_n = k\}$$

The use of stopping times in martingales is given by the following proposition, which is stated without proof.

Proposition 2.3: Let T be a stopping time for a stochastic process $\{X_n\}$, and let $a \wedge b = \min(a,b)$.

1. If $\{X_n\}$ is a martingale, then so is $\{X_{T \wedge n}\}$.
2. If $\{X_n\}$ is a supermartingale, then so is $\{X_{T \wedge n}\}$.
3. If $\{X_n\}$ is a submartingale, then so is $\{X_{T \wedge n}\}$.

Stopping times can be thought of as the time when a given event occurs. If it has the value $T = \infty$, then the event never occurs. For example, in the case of a symmetric random walk, which is a special type of a birth-and-death process and which is known to be a martingale, we may be interested in the first time the value of the process is 6. Then we consider the martingale $\{X_n, n \geq 0\}$ and a random variable T that is defined by

$$T = \begin{cases} \inf_{n \geq 0} \{n \mid X_n = 6\} & \text{if } X_n = 6 \text{ for some } n \in \aleph \\ \infty & \text{otherwise} \end{cases}$$

where $\aleph$ is the set of positive integers. (Random walk is discussed in Chapter 3.)

PROBLEMS

2.1 Suppose $X(t)$ is a Gaussian random process with a mean $E[X(t)]=0$ and auto-correlation function $R_{XX}(\tau)=e^{-|\tau|}$. Assume that the random variable A is defined as follows:

$$A = \int_0^1 X(t)dt$$

Determine the following:
1. $E[A]$
2. σ_A^2

2.2 Three customers A, B, and C simultaneously arrive at a bank with two tellers on duty. The two tellers were idle when the three customers arrived, and A goes directly to one teller, B goes to the other teller, and C waits until either A or B leaves before she can begin receiving service. If the service times provided by the tellers are exponentially distributed with a mean of 4 min, what is the probability that customer A is still in the bank after the other two customers leave?

2.3 Customers arrive at a bank according to a Poisson process with an average rate of six customers per hour. Each arriving customer is either a man with probability p or a woman with probability $1-p$. It was found that in the first 2 h, the average number of men who arrived at the bank was eight. What is the average number of women who arrived over the same period?

2.4 Bob has a pet that requires the light in his apartment to always be on. To achieve this, Bob keeps three light bulbs on with the hope that at least one bulb will be operational when he is not at the apartment. The light bulbs have independent and identically distributed lifetimes T with PDF $f_T(t)=\lambda e^{-\lambda t}$, $\lambda>0$, $t\geq0$.
1. Probabilistically speaking, given that Bob is about to leave the apartment and all three bulbs are working fine, what does he gain by replacing all three bulbs with new ones before he leaves?
2. Suppose X is the random variable that denotes the time until the first bulb fails. What is the PDF of X?
3. Given that Bob is going away for an indefinite period of time and all three bulbs are working fine before he leaves, what is the PDF of Y, the time until the third bulb failure after he leaves?
4. What is the expected value of Y?

2.5 Joe replaced two light bulbs, one of which is rated 60W with an exponentially distributed lifetime whose mean is 200h, and the other is rated 100W with an exponentially distributed lifetime whose mean is 100h.
1. What is the probability that the 60-W bulb fails before the 100-W bulb?
2. What is the mean time until the first of the two bulbs fails?

3. Given that the 60-W bulb has not failed after 300 h, what is the probability that it will last at least another 100 h?

2.6   A five-motor machine can operate properly if at least three of the five motors are functioning. If the lifetime X of each motor has the PDF $f_X(x) = \lambda e^{-\lambda x}$, $\lambda > 0$, $x \geq 0$, and if the lifetimes of the motors are independent, what is the mean of the random variable Y, the time until the machine fails?

2.7   Suzie has two identical personal computers, which she never uses at the same time. She uses one PC at a time, and the other is a backup. If the one she is currently using fails, she turns it off, calls the PC repairman, and turns on the backup PC. The time until either PC fails when it is in use is exponentially distributed with a mean of 50 h. The time between the moment a PC fails until the repairman comes and finishes repairing it is also exponentially distributed with a mean of 3 h. What is the probability that Suzie is idle because neither PC is operational?

2.8   Cars arrive from the northbound section of an intersection in a Poisson manner at the rate of λ_N cars per minute and from the eastbound section in a Poisson manner at the rate of λ_E cars per minute.

1. Given that there is currently no car at the intersection, what is the probability that a northbound car arrives before an eastbound car?

2. Given that there is currently no car at the intersection, what is the probability that the fourth northbound car arrives before the second eastbound car?

2.9   A one-way street has a fork in it, and cars arriving at the fork can either bear right or left. A car arriving at the fork will bear right with probability 0.6 and will bear left with probability 0.4. Cars arrive at the fork in a Poisson manner with a rate of 8 cars per minute.

1. What is the probability that at least four cars bear right at the fork in 3 min?

2. Given that three cars bear right at the fork in 3 min, what is the probability that two cars bear left at the fork in 3 min?

3. Given that 10 cars arrive at the fork in 3 min, what is the probability that four of the cars bear right at the fork?

2.10 Consider the following transition probability matrix:

$$P = \begin{bmatrix} 0.5 & 0.25 & 0.25 \\ 0.3 & 0.3 & 0.4 \\ 0.25 & 0.5 & 0.25 \end{bmatrix}$$

1. Calculate $p_{13}(3)$, $p_{22}(2)$ and $p_{32}(4)$.

2. Calculate $p_{32}(\infty)$.

2.11 A taxicab company has a small fleet of three taxis that operate from the company's station. The time it takes a taxi to take a customer to his or her location and return to the station is exponentially distributed with a mean of $1/\mu$ hours. Customers arrive according to a Poisson process with average rate of λ customers per hour. If a potential customer arrives at the station and finds that no taxi is available, he or she goes to another taxicab company. The taxis always return to the station after dropping off a customer without picking up any new customers on their way back.

1. Give the state-transition-rate diagram of the process.
2. What is the probability that an arriving customer sees exactly one taxi at the station?
3. What is the probability that an arriving customer goes to another taxicab company?

2.12 Let the random variable S_n be defined as follows:

$$
S_n = \begin{cases} 0 & n = 0 \\ \displaystyle\sum_{k=1}^{n} X_k & n \geq 1 \end{cases}
$$

where X_k is the kth outcome of a Bernoulli trial such that $P[X_k = 1] = p$ and $P[X_k = -1] = q = 1 - p$, and the X_k are independent and identically distributed. Consider the process $\{S_n | n = 1, 2, \ldots\}$.

1. For what values of p (relative to q) is $\{S_n\}$ a martingale?
2. For what values of p is $\{S_n\}$ a submartingale?
3. For what values of p is $\{S_n\}$ a supermartingale?

CHAPTER 3

ONE-DIMENSIONAL RANDOM WALK

3.1 INTRODUCTION

A random walk is derived from a sequence of Bernoulli trials. It is used in many fields including thermodynamics, biology, physics, chemistry, and economics where it is used to model fluctuations in the stock market.

Consider a Bernoulli trial in which the probability of success is p and the probability of failure is $1-p$. Assume that the experiment is performed every T time units, and let the random variable X_k denote the outcome of the kth trial. Furthermore, assume that the probability mass function (PMF) of X_k is as follows:

$$p_{X_k}(x) = \begin{cases} p & x = 1 \\ 1 - p & x = -1 \end{cases}$$

That is, the X_k are independent and identically distributed random variables. Finally, let the random variable Y_n be defined as follows:

$$Y_0 = 0$$
$$Y_n = \sum_{k=1}^{n} X_k = Y_{n-1} + X_n \qquad n = 1, 2, \ldots \qquad (3.1)$$

If X_k models a process in which we take a step to the right when the outcome of the kth trial is a success and a step to the left when the outcome is a failure, then

---

Elements of Random Walk and Diffusion Processes, First Edition. Oliver C. Ibe.
© 2013 John Wiley & Sons, Inc. Published 2013 by John Wiley & Sons, Inc.

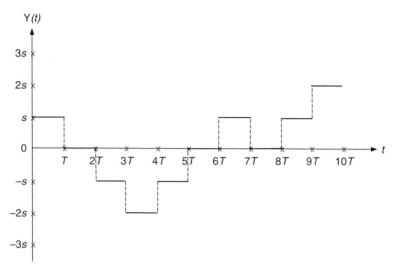

FIGURE 3.1   A sample path of the random walk.

the random variable Y_n represents the location of the process relative to the starting point (or origin) at the end of the nth trial. The resulting trajectory of the process $\{Y_n\}$ as it moves through the x–y plane, where the x coordinate represents the time and the y coordinate represents the location at a given time, is called a one-dimensional *random walk* generated by $\{X_k\}$. If we define the stochastic process $Y(t)=\{Y_n|n\leq t<n+1\}$, then Figure 3.1 shows an example of the sample path of $Y(t)$, where the length of each step is s. It is a staircase with discontinuities at $t=kT$, $k=1, 2, ...$

Suppose that at the end of the nth trial there are exactly k successes. Then there are k steps to the right and $n-k$ steps to the left. Thus,

$$Y(nT) = ks - (n-k)s = (2k-n)s = rs$$

where $r=2k-n$. This implies that $Y(nT)$ is a random variable that assumes values rs, where $r=n$, $n-2$, $n-4$, ..., $-n$. Because the event $\{Y(nT)=rs\}$ is the event $\{k$ successes in n trials$\}$, where $k=(n+r)/2$, we have that

$$P\big[Y(nT) = rs\big] = P\left[\frac{n+r}{2} \text{ successes}\right] = \binom{n}{\frac{n+r}{2}} p^{\frac{n+r}{2}} (1-p)^{\frac{n-r}{2}} \qquad (3.2)$$

Note that $n+r$ must be an even number. Also, because $Y(nT)$ is the sum of n independent Bernoulli random variables, its mean and variance are given as follows:

$$E[Y(nT)] = nE[X_k] = n[ps - (1-p)s] = (2p-1)ns$$
$$E\big[X_k^2\big] = ps^2 + (1-p)s^2 = s^2 \qquad (3.3)$$
$$\mathrm{Var}[Y(nT)] = n\sigma_{X_k}^2 = n\big[s^2 - s^2(2p-1)^2\big] = 4p(1-p)ns^2$$

In the special case where $p=1/2$, $E[Y(nT)]=0$, and $\text{Var}[Y(nT)]=ns^2$. In this case the random walk is called a *symmetric random walk*. If in addition $s=\pm 1$, it is known as a *Bernoulli random walk*.

The one-dimensional Bernoulli random walk possesses the so-called *skip-free property*, which means that to go from state a to state b, the process must pass through all intermediate states because its value can change by at most 1 at each step. This is because a random walk of length n is essentially a system of integers $\{X_0, X_1, \ldots, X_n\}$ such that $X_0=0$ and $|X_i - X_{i-1}|=1$.

3.2 OCCUPANCY PROBABILITY

Let the random variable $Y_n=X_1+X_2+\ldots+X_n$, where the X_k are independent and identically distributed Bernoulli random variables. We would like to find the value of $P[Y_n=m]$, the probability of being at location m at the end of the nth step.

We assume that $X_k=1$ corresponds to a step to the right in the kth trial and $X_k=-1$ corresponds to a step to the left. Suppose that out of the n steps the process takes r steps to the right and, therefore, $n-r$ steps to the left. Thus, we have that $Y_n=m=r-(n-r)=2r-n$, which gives $r=(n+m)/2$. Using the binomial distribution, we obtain

$$P\left[Y_n = m \big| Y_0 = 0\right] = \begin{cases} \dbinom{n}{\dfrac{n+m}{2}} p^{\frac{n+m}{2}} (1-p)^{\frac{n-m}{2}} & n+m \quad \text{even} \\ 0 & n+m \quad \text{odd} \end{cases} \tag{3.4}$$

For the special case of $m=0$ we have that

$$u_n \equiv P\left[Y_n = 0 \big| Y_0 = 0\right] = \begin{cases} \dbinom{n}{\dfrac{n}{2}} \left[p(1-p)\right]^{n/2} & n \quad \text{even} \\ 0 & n \quad \text{odd} \end{cases}$$

which makes sense because this means that the process took as many steps to the right as it did to the left for it to return to the origin. Thus, we may write

$$u_{2n} = \binom{2n}{n}\left[p(1-p)\right]^n \quad n = 0, 1, 2, \ldots \tag{3.5}$$

u_{2n} is the probability of a return to the origin in $2n$ steps. The value of $P[Y_n=m|Y_0=0]$ when n is large can be obtained from the Stirling's approximation as follows:

$$n! \sim (2\pi)^{\frac{1}{2}} n^{n+\frac{1}{2}} e^{-n} \Rightarrow \log(n!) \sim \left(n+\frac{1}{2}\right)\log n - n + \frac{1}{2}\log 2\pi$$

Thus,

$$\lim_{n\to\infty} P\left[Y_n = m \mid Y_0 = 0\right] = \begin{pmatrix} n \\ \dfrac{n+m}{2} \end{pmatrix} p^{\frac{n+m}{2}} (1-p)^{\frac{n-m}{2}}$$

$$= \frac{n!}{\left(\dfrac{n+m}{2}\right)!\left(\dfrac{n-m}{2}\right)!} p^{\frac{n+m}{2}} (1-p)^{\frac{n-m}{2}}$$

$$\approx \frac{\left\{(2\pi)^{\frac{1}{2}} n^{n+\frac{1}{2}} e^{-n}\right\} p^{\frac{n+m}{2}} (1-p)^{\frac{n-m}{2}}}{\left\{(2\pi)^{\frac{1}{2}}\left(\dfrac{n+m}{2}\right)^{\left(\frac{n+m+1}{2}\right)} e^{-\left(\frac{n+m}{2}\right)}\right\}\left\{(2\pi)^{\frac{1}{2}}\left(\dfrac{n-m}{2}\right)^{\left(\frac{n-m+1}{2}\right)} e^{-\left(\frac{n-m}{2}\right)}\right\}}$$

$$= \frac{2^{n+1}\left\{n^{n+\frac{1}{2}}\right\} p^{\frac{n+m}{2}} (1-p)^{\frac{n-m}{2}}}{(2\pi)^{\frac{1}{2}} (n+m)^{(n+m+1)/2} (n-m)^{(n-m+1)/2}} \tag{3.6}$$

Taking logarithm on both sides, we obtain

$$\lim_{n\to\infty} \log\left\{P[Y_n = m \mid Y_0 = 0]\right\}$$

$$= \left(n+\frac{1}{2}\right)\log 2 - \frac{1}{2}\log \pi + \left(n+\frac{1}{2}\right)\log n - \left(\frac{n+m+1}{2}\right)\log\left\{n\left(1+\frac{m}{n}\right)\right\}$$

$$- \left(\frac{n-m+1}{2}\right)\log\left\{n\left(1-\frac{m}{n}\right)\right\} + \left(\frac{n+m}{2}\right)\log p + \left(\frac{n-m}{2}\right)\log q$$

$$= \left(n+\frac{1}{2}\right)\log 2 - \frac{1}{2}\log \pi - \frac{1}{2}\log n - \left(\frac{n+m+1}{2}\right)\log\left\{1+\frac{m}{n}\right\}$$

$$- \left(\frac{n-m+1}{2}\right)\log\left\{1-\frac{m}{n}\right\} + \left(\frac{n+m}{2}\right)\log p + \left(\frac{n-m}{2}\right)\log q$$

where $q = 1-p$. Now, from the Taylor series expansion we know that

$$\log\left(1-\frac{m}{n}\right) = -\frac{m}{n} - \frac{m^2}{2n^2} + O(n^3) \quad -1 \le \frac{m}{n} < 1$$

$$\log\left(1+\frac{m}{n}\right) = \frac{m}{n} - \frac{m^2}{2n^2} + O(n^3) \quad -1 < \frac{m}{n} \le 1$$

Thus,

$$
\lim_{n\to\infty} \log\left\{ P\big[\, Y_n = m \,\big|\, Y_0 = 0 \,\big] \right\} = \left(n + \frac{1}{2} \right) \log 2 - \frac{1}{2}\log\pi - \frac{1}{2}\log n + \left(\frac{n+m}{2} \right) \log p
$$
$$
+ \left(\frac{n-m}{2} \right) \log q - \left(\frac{n+m+1}{2} \right)\left\{ \frac{m}{n} - \frac{m^2}{2n^2} \right\}
$$
$$
+ \left(\frac{n-m+1}{2} \right)\left\{ \frac{m}{n} + \frac{m^2}{2n^2} \right\}
$$
$$
= (n+1)\log 2 - \frac{1}{2}\log 2\pi n - \frac{m^2(n-1)}{2n^2}
$$
$$
+ \left(\frac{n+m}{2} \right)\log p + \left(\frac{n-m}{2} \right)\log q
$$
$$
\approx n\log 2 - \frac{1}{2}\log 2\pi n - \frac{m^2}{2n} + \left(\frac{n+m}{2} \right)\log p + \left(\frac{n-m}{2} \right)\log q
$$

which gives

$$
\lim_{n\to\infty} P\big[\, Y_n = m \,\big|\, Y_0 = 0 \,\big] \approx \left\{ \frac{2^{\,n-\frac{1}{2}}}{\sqrt{n\pi}} \right\} p^{(n+m)/2} q^{(n-m)/2} e^{-m^2/2n}
$$
$$
\approx \left\{ \frac{2^{\,n}}{\sqrt{n\pi}} \right\} p^{(n+m)/2} q^{(n-m)/2} e^{-m^2/2n} \qquad (3.7)
$$
$$
\lim_{n\to\infty} u_{2n} \approx \left\{ \frac{2^{\,n}}{\sqrt{n\pi}} \right\} p^{n/2} q^{n/2}
$$

In the special case where $p = q = 1/2$, we obtain

$$
\lim_{n\to\infty} P\big[\, Y_n = m \,\big|\, Y_0 = 0 \,\big] \approx \left\{ \frac{1}{\sqrt{n\pi}} \right\} e^{-m^2/2n}
$$
$$
\lim_{n\to\infty} u_{2n} \approx \frac{1}{\sqrt{n\pi}}
$$

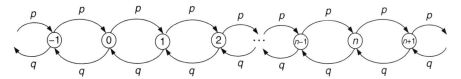

FIGURE 3.2   State transition diagram of a random walk.

3.3 RANDOM WALK AS A MARKOV CHAIN

Let Y_n be as defined earlier. Then

$$P\left[Y_{n+1}=k\,\middle|\,Y_1,Y_2,\ldots,Y_{n-1},Y_n=m\right]=P\left[X_{n+1}=k-m\,\middle|\,Y_1,Y_2,\ldots,Y_{n-1},Y_n=m\right]$$

$$=P\left[X_{n+1}=k-m\,\middle|\,X_1,X_1+X_2,\ldots,\sum_{i=1}^{n-1}X_i,\sum_{i=1}^{n}X_i=m\right]$$

$$=P\left[X_{n+1}=k-m\right]\quad\text{by the independence of }X_{n+1}$$

$$=P\left[X_{n+1}=k-m\,\middle|\,\sum_{i=1}^{n}X_i=m\right]$$

$$=P\left[X_{n+1}=k-m\,\middle|\,Y_n=m\right]$$

$$=P\left[Y_{n+1}=k\,\middle|\,Y_n=m\right]$$

where the third equality is due to the independence of X_{n+1} and the other n random variables. Thus, the future of a random walk depends only on the most recent past outcome. Figure 3.2 shows the state transition diagram of the Markov chain for a one-dimensional random walk.

If with probability 1 a random walker revisits its starting point, the walk is defined to be a *recurrent random walk*; otherwise, it is defined to be nonrecurrent. Later in this chapter we shall show that

$$P\left[Y_n=0\,\middle|\,Y_0=0,\quad n=1,2,\ldots\right]=1-|p-q|$$

Thus, a random walk is recurrent only if it is a symmetric random walk; that is, only if $p=q=1/2$.

3.4 SYMMETRIC RANDOM WALK AS A MARTINGALE

Consider the random walk $\{Y_n\,|\,n=0,1,2,\ldots\}$ where $Y_0=0$ and

$$Y_n=\sum_{k=1}^{n}X_k=Y_{n-1}+X_n\quad n=1,2,\ldots$$

Assume that $P[X_k=1]=p$ and $P[X_k=-1]=1-p$. Then

$$
\begin{aligned}
E\left[Y_n | Y_0, Y_1, \ldots, Y_k\right] &= E\left[Y_n - Y_k + Y_k | Y_0, Y_1, \ldots, Y_k\right] \\
&= E\left[Y_n - Y_k | Y_0, Y_1, \ldots, Y_k\right] + E\left[Y_k | Y_0, Y_1, \ldots, Y_k\right] \\
&= E\left[Y_n - Y_k | Y_0, Y_1, \ldots, Y_k\right] + Y_k \\
&= E\left[\sum_{j=k+1}^{n} \left(X_j | Y_0, Y_1, \ldots, Y_k\right)\right] + Y_k \\
&= \sum_{j=k+1}^{n} E\left[X_j | X_1, \ldots, X_k\right] + Y_k = \sum_{j=k+1}^{n} E\left[X_j\right] + Y_k \\
&= (n-k)(2p-1) + Y_k
\end{aligned}
$$

The last equality follows from the fact that $E[X_k]=(1)p+(-1)(1-p)=2p-1$. Now, for a symmetric random walk, $p=1/2$ and the result becomes

$$
E\left[Y_n | Y_0, Y_1, \ldots, Y_k\right] = Y_k
$$

Thus, a symmetric random walk is a martingale, but a nonsymmetric random walk is not a martingale.

3.5   RANDOM WALK WITH BARRIERS

The random walk previously described assumes that the process can continue forever; in other words, it is unbounded. Sometimes the walker cannot go outside some defined boundaries, in which case the walk is said to be a restricted random walk and the boundaries are called *barriers*. These barriers can impose different characteristics on the walk process. For example, they can be *reflecting barriers*, which means that on hitting them the walk turns around and continues. They can also be *absorbing barriers*, which means that the walk ends when a barrier is hit. Figure 3.3 shows the different types of barriers where it is assumed that the number of states is finite.

When absorbing barriers are present we obtain an absorbing Markov chain. One example of the random walk with barriers is the gambler's ruin problem that is discussed later in this chapter.

3.6   MEAN-SQUARE DISPLACEMENT

The distance of the walker from the origin after N steps is given by

$$
Y_N = X_1 + X_2 + \cdots + X_N
$$

(a)

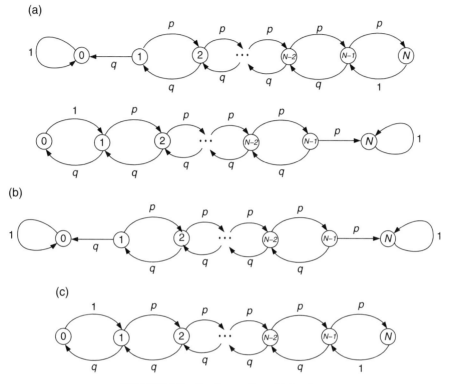

(b)

(c)

FIGURE 3.3   Random walks with barriers.

The expected value of Y_N is given by

$$E[Y_N] = E\left[\sum_{i=1}^{N} X_i\right] = \sum_{i=1}^{N} E[X_i] = N\left[p(\delta) + (1-p)(-\delta)\right]$$
$$= N\left[2p\delta - \delta\right]$$

where δ is the step size. Thus, for a symmetric walk, $E[Y_N]=0$, as stated earlier. To find the spread of the displacement from the origin we compute $E\left[Y_N^2\right]$, which is given by

$$E\left[Y_N^2\right] = E\left[\left(X_1 + X_2 + \cdots + X_N\right)^2\right]$$
$$= E\left[X_1^2\right] + E\left[X_2^2\right] + \cdots + E\left[X_N^2\right] + 2\left\{E[X_1 X_2] + E[X_1 X_3] + \cdots + E[X_1 X_N]\right\}$$
$$+ 2\left\{E[X_2 X_3] + E[X_2 X_4] + \cdots + E[X_2 X_N]\right\} + \cdots + 2E[X_{N-1} X_N]$$

Since the X_n are independent random variables, $E[X_i X_j] = E[X_i]E[X_j] = (2p\delta - \delta)^2$. Thus, for an asymmetric random walk we have that

$$E\left[Y_N^2\right] = NE[X_1^2] + 2(2p\delta - \delta)^2 \left\{(N-1) + (N-2) + \cdots + 2 + 1\right\}$$
$$= N\left\{p\delta^2 + (1-p)\delta^2\right\} + N(N-1)(2p\delta - \delta)^2 \qquad (3.8)$$
$$= N\delta^2 + N(N-1)(2p\delta - \delta)^2$$

For a symmetric random walk we have that $E[X_iX_j] = (2p\delta - \delta)^2 = 0$, which means that

$$E\left[Y_N^2\right] = N\delta^2 \qquad (3.9)$$

Let $t = N\tau$ be the time it takes to complete the N steps; then $N = t/\tau$. We define a diffusion constant $D = \delta^2/2\tau$, which means that Equation 3.9 becomes

$$E\left[Y_N^2(t)\right] = 2Dt \qquad (3.10)$$

3.7 GAMBLER'S RUIN

Consider the following random walk with two absorbing barriers, which is generally referred to as the gambler's ruin. Suppose a gambler plays a sequence of independent games against an opponent. He starts out with \$$k$, and in each game he wins \$1 with probability p and loses \$1 with probability $q = 1 - p$. When $p > q$, the game is advantageous to the gambler either because he is more skilled than his opponent or the rules of the game favor him. If $p = q$, the game is fair; if $p < q$, the game is disadvantageous to the gambler.

3.7.1 Ruin Probability

Assume that the gambler stops when he has a total of \$0 or \$$N$. In the latter case, he has additional \$$(N-k)$ over his initial \$$k$. (Another way to express this is that he plays against an opponent who starts out with \$$(N-k)$, and the game stops when either player has lost all his money.) We are interested in computing the probability r_k that the player will be ruined (or he has lost all his money) after starting with \$$k$. The state-transition diagram of the process is illustrated in Figure 3.3**b** where the states represent the total amount the gambler currently has.

To solve the problem, we note that at the end of the first game, the player will have the sum of \$$(k+1)$ if he wins the game (with probability p) and the sum of \$$(k-1)$ if he loses the game (with probability q). Thus, if he wins the first game, the probability that he will eventually be ruined is r_{k+1}; if he loses his first game, the probability that he will be ruined is r_{k-1}. There are two boundary conditions in this problem. First, $r_0 = 1$ because he cannot gamble when he has no money. Second, $r_N = 0$ because he cannot be ruined since his opponent is already ruined. Thus, we obtain the following difference equation:

$$r_k = qr_{k-1} + pr_{k+1} \qquad 0 < k < N \qquad (3.11)$$

Because $p+q=1$, Equation 3.11 becomes

$$(p+q)r_k = qr_{k-1} + pr_{k+1} \quad 0<k<N$$

which we can write as

$$p(r_{k+1} - r_k) = q(r_k - r_{k-1})$$

From this we obtain the following:

$$r_{k+1} - r_k = \left(\frac{q}{p}\right)(r_k - r_{k-1}) \quad 0<k<N \tag{3.12}$$

We observe that

$$r_2 - r_1 = \left(\frac{q}{p}\right)(r_1 - r_0) = \left(\frac{q}{p}\right)(r_1 - 1)$$

$$r_3 - r_2 = \left(\frac{q}{p}\right)(r_2 - r_1) = \left(\frac{q}{p}\right)^2 (r_1 - 1)$$

$$r_4 - r_3 = \left(\frac{q}{p}\right)(r_3 - r_2) = \left(\frac{q}{p}\right)^3 (r_1 - 1)$$

and so on; thus, we obtain the following:

$$r_{k+1} - r_k = \left(\frac{q}{p}\right)^k (r_1 - 1) \quad 0<k<N \tag{3.13}$$

Now,

$$r_k - 1 = r_k - r_0 = (r_k - r_{k-1}) + (r_{k-1} - r_{k-2}) + \cdots + (r_2 - r_1) + (r_1 - 1)$$

$$= \left[\left(\frac{q}{p}\right)^{k-1} + \left(\frac{q}{p}\right)^{k-2} + \cdots + \left(\frac{q}{p}\right) + 1\right](r_1 - 1)$$

$$= \begin{cases} \dfrac{1-(q/p)^k}{1-(q/p)}(r_1 - 1) & p \neq q \\ k(r_1 - 1) & p = q \end{cases}$$

The boundary condition that $r_N = 0$ implies that

$$r_1 = \begin{cases} 1 - \dfrac{1-(q/p)}{1-(q/p)^N} & p \neq q \\ 1 - \dfrac{1}{N} & p = q \end{cases}$$

Thus,

$$
r_k = \begin{cases} 1 - \dfrac{1-(q/p)^k}{1-(q/p)^N} = \dfrac{(q/p)^k - (q/p)^N}{1-(q/p)^N} & p \ne q \\[4mm] 1 - \dfrac{k}{N} & p = q \end{cases}
\tag{3.14}
$$

3.7.2   Alternative Derivation of Ruin Probability

An alternative method of deriving r_k is to try a solution of the form $r_k = z^k$ for the equation

$$
r_k = q r_{k-1} + p r_{k+1} \quad 0 < k < N
$$

This gives

$$
p z^{k+1} + q z^{k-1} = z^k \Rightarrow p z^2 + q = z \Rightarrow p z^2 - z + q = 0
$$

Thus,

$$
z = \frac{1 \pm \sqrt{1 - 4pq}}{2p} = \frac{1 \pm \sqrt{1 - 4p(1-p)}}{2p} = \frac{1 \pm \sqrt{1 - 4p + 4p^2}}{2p}
$$

$$
= \frac{1 \pm \sqrt{(1-2p)^2}}{2p} = \frac{1 \pm (1-2p)}{2p}
$$

The solutions are the two roots $z=1$ and $z=(1-p)/p=q/p$. Thus, we have that

$$
r_k = a(1)^k + b\left(\frac{q}{p}\right)^k = a + b\left(\frac{q}{p}\right)^k
$$

where a and b are arbitrary constants. If we assume that $p \ne q$, we have that $r_0 = 1 = a + b \Rightarrow a = 1 - b$. Similarly, the condition $r_N = 0 = a + b(q/p)^N$ gives $a = -b(q/p)^N$. Solving the two equations $a = 1 - b = -b(q/p)^N$ gives

$$
b = \frac{1}{1-(q/p)^N}, \quad a = 1 - b = \frac{-(q/p)^N}{1-(q/p)^N}
$$

which gives the result in Equation 3.14 for the case of $p \ne q$. Similarly, when $p=q$, we obtain the double root $z=1$, which gives

$$
r_k = a(1)^k + bk(1)^k = a + bk
$$

The condition $r_0 = 1 \Rightarrow a = 1$, and the condition $r_N = 0 \Rightarrow b = -a/N = -1/N$, which gives the result in Equation 3.14 for the case of $p=q$.

3.7.3   Duration of a Game

Let d_k denote the expected time until a gambler who starts with \$$k$ is ruined. Clearly, $d_0=d_N=0$. If the gambler wins the first game, the expected duration of the game is $d_{k+1}+1$; if he loses the first game, the expected duration of the game is $d_{k-1}+1$. Thus, d_k satisfies the following difference equation:

$$d_k = q(d_{k-1}+1) + p(d_{k+1}+1) = 1 + qd_{k-1} + pd_{k+1} \quad 0<k<N$$

Because $p+q=1$, we can rewrite the aforementioned difference equation as follows:

$$pd_{k+1} - pd_k - qd_k + qd_{k-1} + 1 = 0 = p(d_{k+1}-d_k) - q(d_k-d_{k-1}) + 1 \quad 0<k<N$$

That is,

$$p(d_{k+1}-d_k) = q(d_k-d_{k-1}) - 1 \quad 0<k<N$$

Let $m_k=d_k-d_{k-1}$. Then we have that

$$pm_{k+1} = qm_k - 1$$

Solving the preceding equation iteratively, we obtain

$$pm_2 = qm_1 - 1 \Rightarrow m_2 = \frac{q}{p}m_1 - \frac{1}{p}$$

$$pm_3 = qm_2 - 1 \Rightarrow m_3 = \frac{q}{p}m_2 - \frac{1}{p} = \left(\frac{q}{p}\right)^2 m_1 - \frac{1}{p}\left\{1+\frac{q}{p}\right\}$$

$$pm_4 = qm_3 - 1 \Rightarrow m_4 = \frac{q}{p}m_3 - \frac{1}{p} = \left(\frac{q}{p}\right)^3 m_1 - \frac{1}{p}\left\{1+\frac{q}{p}+\left(\frac{q}{p}\right)^2\right\}$$

$$pm_5 = qm_4 - 1 \Rightarrow m_5 = \frac{q}{p}m_4 - \frac{1}{p} = \left(\frac{q}{p}\right)^4 m_1 - \frac{1}{p}\left\{1+\frac{q}{p}+\left(\frac{q}{p}\right)^2+\left(\frac{q}{p}\right)^3\right\}$$

Thus, in general, we have that

$$m_k = \left(\frac{q}{p}\right)^{k-1} m_1 - \frac{1}{p}\left\{1+\frac{q}{p}+\left(\frac{q}{p}\right)^2+\left(\frac{q}{p}\right)^3+\cdots+\left(\frac{q}{p}\right)^{k-2}\right\}$$

$$= \left(\frac{q}{p}\right)^{k-1} m_1 - \frac{1}{p}\sum_{j=0}^{k-2}\left(\frac{q}{p}\right)^j$$

Now, $m_1=d_1-d_0=d_1$. Also, $m_2=d_2-d_1=d_2-m_1\Rightarrow d_2=m_2+m_1$. Similarly, $m_3=d_3-d_2=d_3-m_2-m_1\Rightarrow d_3=m_3+m_2+m_1$. Thus, in general,

$$d_k = \sum_{j=1}^{k} m_j = \sum_{j=1}^{k} \left\{ \left(\frac{q}{p}\right)^{j-1} m_1 - \frac{1}{p} \sum_{i=0}^{j-2} \left(\frac{q}{p}\right)^i \right\}$$

$$= \begin{cases} \dfrac{1-(q/p)^k}{1-(q/p)} \left\{ d_1 + \dfrac{1}{p-q} \right\} - \dfrac{k}{p-q} & p \neq q \\[3mm] k\{d_1 - (k-1)\} & p = q \end{cases}$$

Because $d_N=0$, we have that

$$d_1 = \begin{cases} \dfrac{1-(q/p)}{1-(q/p)^N} \left\{ \dfrac{N}{p-q} \right\} - \dfrac{1}{p-q} & p \neq q \\[3mm] N-1 & p = q \end{cases} \tag{3.15}$$

Thus, we finally obtain

$$d_k = \begin{cases} \dfrac{1-(q/p)^k}{1-(q/p)} \left\{ d_1 + \dfrac{1}{p-q} \right\} - \dfrac{k}{p-q} & p \neq q \\[3mm] k(N-k) & p = q \end{cases} \tag{3.16}$$

3.8   RANDOM WALK WITH STAY

Consider a random walk in which the walker moves to the right with probability p, to the left with probability q or remains in the current position with probability $v=1-p-q$. This is called a random walk with stay or *random walk with hesitation*. It can be used to model a gambling problem in which a round of the game can result in a tie with probability v if each round is independent of other rounds. The state-transition diagram for the process is shown in Figure 3.4.

We use r_k to denote the probability that the player will be ruined after starting in state k. As in the classical gambler's ruin problem without ties, $r_0=1$ and $r_N=0$. Thus, we obtain the following difference equation:

$$r_k = qr_{k-1} + (1-p-q)r_k + pr_{k+1} \quad 0 < k < N$$

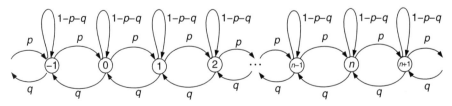

FIGURE 3.4   State-transition diagram for the gambler's ruin problem with stay.

If we assume a solution of the form $r_k = z^k$, the equation becomes

$$z = pz^2 + (1 - p - q)z + q \Rightarrow pz^2 - (p + q)z + q = 0$$

Thus,

$$z = \frac{(p+q) \pm \sqrt{(p+q)^2 - 4pq}}{2p} = \frac{(p+q) \pm \sqrt{(p-q)^2}}{2p} = \frac{(p+q) \pm (p-q)}{2p}$$

That is, $z = 1$ and $z = q/p$, which are the solutions for the classical ruin problem, except that in this case, $p + q \neq 1$.

3.9   FIRST RETURN TO THE ORIGIN

Let T_0 denote the time until a random walk returns to the origin for the first time. That is, given that $Y_n = Y_{n-1} + X_n$, $Y_0 = 0$, $n = 1, 2, \ldots$, we define

$$T_0 = \min\{n \geq 1 : Y_n = 0\} \tag{3.17}$$

as the first return to 0. Its PMF

$$p_{T_0}(n) = P[T_0 = n] = P[Y_1 \neq 0, Y_2 \neq 0, \ldots, Y_{n-1} \neq 0, Y_n = 0]$$

is given by the following theorem:

Theorem 3.1: The z-transform of the probability mass function of the first return to 0 is given by

$$G_{T_0}(z) = 1 - \left(1 - 4pqz^2\right)^{1/2}$$

Proof: Let $p_0(n) = P[Y_n = 0]$ be the probability that the process is at the origin after n steps. Let A be the event that $Y_n = 0$, and let B_k be the event that the first return to the origin occurs at the kth step. Because the B_k are mutually exclusive events, we have that

$$P[A] = \sum_{k=1}^{n} P[A|B_k] P[B_k] \quad n \geq 1$$

Now, $P[B_k] = p_{T_0}(k)$ and $P[A|B_k] = p_0(n-k)$, which means that

$$p_0(n) = \sum_{k=1}^{n} p_0(n-k) p_{T_0}(k) \quad n \geq 1$$

Let the z-transform of $p_0(n)$ be $G_{Y_n}(z)$. As the z-transform of $p_{T_0}(k)$ is given by $G_{T_0}(z)$, the preceding equation becomes

$$G_{Y_n}(z) = \sum_{n=0}^{\infty} p_0(n)z^n = p_0(0) + \sum_{n=1}^{\infty}\sum_{k=1}^{n} p_0(n-k)p_{T_0}(k)z^n$$
$$= 1 + G_{Y_n}(z)G_{T_0}(z)$$

where the last equality follows from the fact that $p_0(0)=1$. Thus, we obtain

$$G_{T_0}(z) = 1 - \frac{1}{G_{Y_n}(z)}$$

To obtain $P[Y_n=0]$, we must move an equal number of steps to the right and to the left. That is,

$$p_0(n) = \begin{cases} \begin{pmatrix} n \\ \dfrac{n}{2} \end{pmatrix}(pq)^{n/2} & n \text{ even} \\ 0 & \text{otherwise} \end{cases}$$

Thus, we have that

$$G_{Y_n}(z) = \sum_{n=0}^{\infty} \begin{pmatrix} 2n \\ n \end{pmatrix}(pq)^n z^{2n} = \sum_{n=0}^{\infty} \begin{pmatrix} 2n \\ n \end{pmatrix}(pqz^2)^n = (1-4pqz^2)^{-1/2}$$

where the last equality follows from the identity (Wilf 1990)

$$\sum_{n=0}^{\infty} \begin{pmatrix} 2n \\ n \end{pmatrix}x^n = \frac{1}{\sqrt{1-4x}}$$

From this we obtain

$$G_{T_0}(z) = 1 - (1-4pqz^2)^{1/2}$$

which completes the proof.

By expanding $G_{T_0}(z)$, as a power series we obtain the distribution of T_0 as follows. We know that according to the binomial theorem

$$(1+a)^n = \sum_{k=0}^{n} \begin{pmatrix} n \\ k \end{pmatrix}a^k$$

If we define $\begin{pmatrix} n \\ k \end{pmatrix}$ to be equal to 0 when $k>n$, then we can omit the upper limit in the sum and write the quantity $(1-a)^n$ as follows:

$$(1-a)^n = \sum_{k\geq 0} \begin{pmatrix} n \\ k \end{pmatrix}(-a)^k$$

Thus,

$$G_{T_0}(z) = 1 - (1 - 4pqz^2)^{1/2} = 1 - \sum_{k \geq 0} \binom{1/2}{k}(-4pqz^2)^k = \sum_{k \geq 1} \binom{1/2}{k}(-1)^{k-1}(4pq)^k z^{2k}$$

which implies that

$$p_{T_0}(2k) = (-1)^{k-1}\binom{1/2}{k}(4pq)^k \quad k = 1, 2, \dots \tag{3.18}$$

If we define v_0 as the probability that the process ever returns to the origin, then we have that

$$v_0 = \sum_{n=1}^{\infty} p_{T_0}(n) = G_{T_0}(1) = P[T_0 < \infty] = 1 - (1 - 4pq)^{1/2}$$

Because $p + q = 1$, we have that

$$(1 - 4pq)^{1/2} = (1 - 4p\{1 - p\})^{1/2} = (1 - 4p + 4p^2)^{1/2} = \{(1 - 2p)^2\}^{1/2} = |1 - 2p|$$

which means that

$$v_0 = 1 - |1 - 2p| = 1 - |p - q| \tag{3.19}$$

For the symmetric random walk (i.e., $p = q = 1/2$), we have that $v_0 = 1$ and

$$G_{T_0}(z) = 1 - (1 - z^2)^{1/2}$$

Because $dG_{T_0}(z)/dz \,|_{z=1} = \infty$, we have that $E[T_0] = \infty$ for a symmetric random walk.

3.10 FIRST PASSAGE TIMES FOR SYMMETRIC RANDOM WALK

Let T_{ij} denote the time a symmetric random walk takes to reach the state j for the first time, given that is started in state i. Thus, T_{ij} is referred to as the first passage time of the random walk from i to j; that is,

$$T_{ij} = \min\{n \geq 0 : Y_0 = i, Y_n = j\}$$

Let the PMF of T_{ij} be $p_{T_{ij}}(k) = P[T_{ij} = k]$. We consider two methods of obtaining $p_{T_{ij}}(k)$. The first method is via the generating function, and the second method is via the reflection principle.

3.10.1 First Passage Time via the Generating Function

Consider T_{01}. In the first step, the walker can reach state 1 (i.e., $X_1 = 1$) with probability 1/2 or state −1 (i.e., $X_1 = -1$) with probability 1/2. If $X_1 = 1$, then $T_{01} = 1$. On the other hand, if $X_1 = -1$, then the walker will first reach 0 and then reach 1. That is, $T_{01} = 1 + T_{-1,1}$. But $T_{-1,1} = T_{-1,0} + T_{01}$ since it has to pass through 0 to get to 1. Because of Markov property, $T_{-1,0}$ and T_{01} have the same distribution and they are independent. Thus,

$$
\begin{aligned}
G_{T_{01}}(z) &= E\left[z^{T_{01}}\right] = E\left[z^{T_{01}} \,\middle|\, X_1 = 1\right] P[X_1 = 1] + E\left[z^{T_{01}} \,\middle|\, X_1 = -1\right] P[X_1 = -1] \\
&= E\left[z^1 \,\middle|\, X_1 = 1\right] P[X_1 = 1] + E\left[z^{1+T_{-1,0}+T_{01}} \,\middle|\, \right] P[X_1 = -1] \\
&= zP[X_1 = 1] + zE\left[z^{T_{-1,0}+T_{01}}\right] P[X_1 = -1] \\
&= zP[X_1 = 1] + zE\left[z^{T_{-1,0}}\right] E\left[z^{T_{01}}\right] P[X_1 = -1] \\
&= \frac{z}{2} + \frac{z}{2}\left\{G_{T_{01}}(z)\right\}^2
\end{aligned}
$$

Solving the quadratic equation $(z/2)\{G_{T_{01}}(z)\}^2 - G_{T_{01}}(z) + (z/2) = 0$ we obtain

$$
G_{T_{01}}(z) = \frac{1 \pm \sqrt{1 - 4\left(\frac{1}{2}z\right)\left(\frac{1}{2}z\right)}}{z} = \frac{1 \pm \sqrt{1 - z^2}}{z}
$$

Because $G_{T_{01}}(z)$ must take values between 0 and 1 when $0 < z < 1$, we must take the negative part of the $\pm$, which gives

$$
G_{T_{01}}(z) = \frac{1 - \sqrt{1 - z^2}}{z}
$$

From Newton's binomial formula,

$$
\sqrt{1 - z^2} = \sum_{k=0}^{\infty} \binom{1/2}{k} (-z^2)^k = \sum_{k=0}^{\infty} (-1)^k \binom{1/2}{k} (z^2)^k \tag{3.20}
$$

From this it can be shown that

$$
P[T_{01} = 2k - 1] = (-1)^{k-1} \binom{1/2}{k}
$$

Now, T_{0j} is the sum of T_{01}, T_{12}, T_{23}, ..., $T_{j-1,j}$ which are independent and identically distributed; that is,

$$
T_{0j} = T_{01} + T_{12} + \cdots + T_{j-1,j}
$$

Thus, the z-transform of T_{0j} is

$$
G_{T_{0j}}(z) = E\left[z^{T_{0j}}\right] = E\left[z^{T_{01}+T_{12}+\cdots+T_{j-1,j}}\right] = \left\{G_{T_{01}}(z)\right\}^j \tag{3.21}
$$

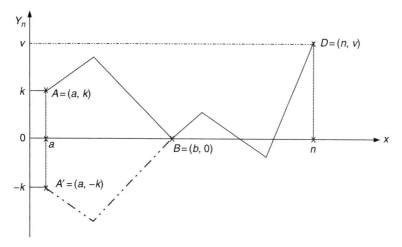

FIGURE 3.5   Illustration of the reflection principle.

Similarly,

$$G_{T_{ij}}(z) = \{G_{T_{01}}(z)\}^{|j-i|} \tag{3.22}$$

3.10.2   First Passage Time via the Reflection Principle

The computation of the probability of an event associated with a random walk is essentially the counting of the number of paths that define that event. These probabilities can often be derived from the *reflection principle,* which states as follows:

Reflection Principle: Let k, $v > 0$. Any path from $A = (a, k)$ to $D = (n, v)$ that touches or crosses the x-axis in between, corresponds to a path from $A' = (a, -k)$ to $D = (n, v)$.

Thus, the x-axis can be thought of as a mirror that casts a "shadow path" of the original path by reflecting it on this mirror until it hits the x-axis for the first time. After the first time the original path hits the x-axis at $B = (b, 0)$, the shadow path is exactly the same as the original path. In Figure 3.5, the segment $A'B$ is the shadow path of the original segment AB. After B, the two segments converge and continue as one path to D. Thus, the number of paths from A to D that touch or cross the x-axis is the same as the number of paths from A' to D.

Let $N_{a,n}(k, v)$ denote the number of possible paths between the points (a, k) and (n, v). (A guide to understanding this parameter is that the subscript indicates the starting and ending times, while the argument indicates the initial and final positions.) Then $N_{a,n}(k, v)$ can be computed as follows. Let a path consist of m steps to the right and l steps to the left. Thus, the total number of steps is $m + l = n - a$, and the difference between the number of rightward steps and the leftward steps is $m - l = v - k$. From this we obtain

$$m = \frac{1}{2}\{n - a + v - k\} \tag{3.23}$$

Because $N_{a,n}(k, v)$ can be defined as the number of "successes," m, in $m+l=n-a$ binomial trials, we have that

$$N_{a,n}(k, v) = \binom{n-a}{m} \qquad (3.24)$$

where m is as defined in Equation 3.23.

Consider the event $\{Y_n = x | Y_0 = 0\}$; that is, the position of the walker after n steps is x, given that he started at the origin. In this case we have that $a=0, k=0, v=x$. Thus, $m=(n-x)/2$ and

$$N_{0,n}(0, x) = \binom{b-a}{m} = \binom{n}{\frac{n+x}{2}} = \binom{n}{\frac{n-x}{2}}$$

where $(n+x)/2$ is an integer. Thus, if p is the probability of a step of 1 and q is the probability of a step of -1,

$$P[Y_n = x] = N_{0,x}(0, x)p^{(n+x)/2}q^{(n-x)/2} = \binom{n}{\frac{n+x}{2}} p^{(n+x)/2}q^{(n-x)/2}$$

as we derived earlier in the chapter.

According to the reflection principle, the number of paths from A to N that touch or cross the time axis is equal to the number of paths from A' to D; that is, if we denote the number of paths that touch or cross the time axis in Figure 3.5 by $N^1_{a,n}(k, v)$, then

$$N^1_{a,n}(k, v) = N_{a,n}(-k, v)$$

If we assume that k and v are positive numbers as shown in Figure 3.5, then the number of paths from (a, k) to (n, v) that do not touch or intersect the time, as denoted by $N^0_{a,n}(k, v)$ is the complement of the number of paths, $N^1_{a,n}(k, v)$, which touch or intersect the time axis. Thus,

$$N^0_{a,n}(k,v) = N_{a,n}(k, v) - N^1_{a,n}(k, v) = N_{a,n}(k,v) - N_{a,n}(-k, v)$$

To apply this principle to the first passage time problem, we proceed as follows. Consider a random walk that starts at $A=(0, 0)$, crosses or touches a line $y>v$, and then ends at $D=(n, v)$. This is illustrated in Figure 3.6. From our earlier discussion, the total number of paths between $(0, 0)$ and (n, x) is $N_{0,n}(0, v)$.

Consider a reflection on the line $Y_n = y$, as shown in Figure 3.6. Assume that the last point at which the path from $(0, 0)$ to (n, v) intersects this line is the point $B=(k, y)$. Then the reflection of the path from $(0, 0)$ to (n, v) on this line form the point B is shown

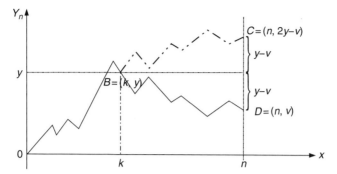

FIGURE 3.6   Reflection principle illustrated for first passage time.

in dotted line. The terminal point of this line is $C=(n, 2y-v)$. According to the reflection principle, the number of paths from A to D that intersect or touch the line $Y_n=y$ is

$$N_{0,n}^1(0, v) = N_{0,n}(0, 2y-v)$$

To compute the first passage time, we note that for the walker to be at the point v at time n, he must be either at the point $v-1$ at time $n-1$ or at the point $v+1$ at time $n-1$. Since his first time of reaching the point v is n, we conclude that he must be at $v-1$ at time $n-1$. Thus, the number of paths from A to D that do not touch or cross $Y=v$ before time $t-1$, $N_{0,n}^0(0, v)$, is

$$
\begin{aligned}
N_{0,n}^0(0, v) &= N_{0,n}(0, v) - N_{0,n}^1(0, v) \\
&= N_{0,n-1}(0, v-1) - N_{0,n-1}(0, 2v-(v-1)) \\
&= N_{0,n-1}(0, v-1) - N_{0,n-1}(0, v+1) \\
&= \binom{n-1}{\frac{n+v-2}{2}} - \binom{n-1}{\frac{n+v}{2}} = \frac{(n-1)!}{\left(\frac{n+v-2}{2}\right)!\left(\frac{n-v}{2}\right)!} - \frac{(n-1)!}{\left(\frac{n+v}{2}\right)!\left(\frac{n-v-2}{2}\right)!} \\
&= \frac{(n-1)!}{\left(\frac{n+v}{2}-1\right)!\left(\frac{n-v}{2}\right)!} - \frac{(n-1)!}{\left(\frac{n+v}{2}\right)!\left(\frac{n-v}{2}-1\right)!} \\
&= \frac{(n-1)!}{\left(\frac{n+v}{2}\right)!\left(\frac{n-v}{2}\right)!}\left\{\frac{n+v}{2} - \frac{n-v}{2}\right\} \\
&= \left\{\frac{v}{n}\right\}\frac{n!}{\left(\frac{n+v}{2}\right)!\left(\frac{n-v}{2}\right)!} = \left\{\frac{v}{n}\right\}\binom{n}{\frac{n+v}{2}} \\
&= \frac{v}{n}N_{0,n}(0, v)
\end{aligned}
$$

where $N_{0,n}(0, v)$ is the total number of paths from A to D. Thus, the probability that the first passage time from A to D occurs in n steps is

$$p_{T_v}(n) = P[T_v = n] = \frac{v}{n} \binom{n}{\frac{n+v}{2}} p^{\frac{n+v}{2}} (1-p)^{\frac{n-v}{2}}$$

Because a similar result can be obtained when $v < 0$, we have that

$$p_{T_v}(n) = P[T_v = n] = \frac{|v|}{n} \binom{n}{\frac{n+v}{2}} p^{\frac{n+v}{2}} (1-p)^{\frac{n-v}{2}} \tag{3.25}$$

Note that $n+v$ must be an even number. Also, recall that the probability that the walker is at location v after n steps is given by

$$P[Y_n = v] = \binom{n}{\frac{n+v}{2}} p^{\frac{n+v}{2}} (1-p)^{\frac{n-v}{2}}$$

Thus, the PMF for the first passage time can be written as follows:

$$p_{T_v}(n) = P[T_v = n] = \frac{|v|}{n} P[Y_n = v] \tag{3.26}$$

3.10.3 Hitting Time and the Reflection Principle

One way to state the reflection principle is that if point m is hit at time $T_m < n$, then it is equally likely, by symmetry, that Y_n will be either above or below m. Alternatively, for every sample path starting at T_m and ending above m at time n, there exists a symmetrical path starting at T_m and ending below m at time n. That is,

$$P[Y_n \geq m \,|\, T_m \leq n] = P[Y_n \leq m \,|\, T_m \leq n] = 0.5$$

From the law of total probability we have that

$$P[Y_n \geq m] = P[Y_n \geq m \,|\, T_m \leq n]P[T_m \leq n] + P[Y_n \geq m \,|\, T_m > n]P[T_m > n]$$

But $P[Y_n \ge m | T_m > n] = 0$. Thus we have that

$$P[Y_n \ge m] = 0.5 P[T_m \le n] \Rightarrow P[T_m \le n] = 2P[Y_n \ge m] \qquad (3.27)$$

3.11 THE BALLOT PROBLEM AND THE REFLECTION PRINCIPLE

Consider Figure 3.5 and assume that $n > a \ge 0$ and $k > 0$, $v > 0$. According to the reflection principle, the number of paths from A to D that touch or cross the time axis is equal to the number of paths from A' to D. To obtain a path $(Y_0=0, Y_1, \ldots, Y_k=x)$ from $(0, 0)$ to (k, x) such that $Y_1 > 0, \ldots, Y_k > 0$, we must have $Y_1 = 1$. Thus, the probability that a path from $(0, 0)$ to (n, x) does not touch or cross the time axis is the ratio of the number of paths from $(1,1)$ to (n, x) that do not touch or cross the time axis, $N^0_{1,n}(1, x)$, to the total number of paths from $(1, 1)$ to (n, x). But from the reflection principle, the number of paths that touch or cross the time axis from $(1, 1)$ and end at (n, x), $N^1_{1,n}(1, x)$, is the number of paths from $(1, -1)$ to (n, x). That is,

$$N^1_{1,n}(1, x) = N_{1,n}(-1, x)$$

Thus,

$$N^0_{1,n}(1,x) = N_{1,n}(1,x) - N^1_{1,n}(1,x) = N_{1,n}(1,x) - N_{1,n}(-1,x)$$

$$= \binom{n-1}{\dfrac{n+x-2}{2}} - \binom{n-1}{\dfrac{n+x}{2}}$$

$$= \dfrac{(n-1)!}{\left(\dfrac{n-x}{2}\right)!\left(\dfrac{n+x}{2}-1\right)!} - \dfrac{(n-1)!}{\left(\dfrac{n-x}{2}-1\right)!\left(\dfrac{n+x}{2}\right)!}$$

$$= \dfrac{(n-1)!}{\left(\dfrac{n-x}{2}\right)!\left(\dfrac{n+x}{2}\right)!}\left\{\dfrac{n+x}{2} - \dfrac{n-x}{2}\right\} \qquad (3.28)$$

$$= \dfrac{n!}{n\left(\dfrac{n-x}{2}\right)!\left(\dfrac{n+x}{2}\right)!}\{x\} = \left\{\dfrac{x}{n}\right\}\dfrac{n!}{\left(\dfrac{n-x}{2}\right)!\left(\dfrac{n+x}{2}\right)!}$$

$$= \left\{\dfrac{x}{n}\right\}\binom{n}{\dfrac{n+x}{2}}$$

$$= \dfrac{x}{n}N_{0,n}(0,x)$$

Therefore, the probability of a path from $(0, 0)$ to (k, x) not to touch or cross the time axis is

$$\frac{N_{1,n}^0(1, x)}{N_{0,n}(0, x)} = \frac{x}{n}$$

This result can be used to solve the ballot problem, which can be defined as follows. Consider an election involving only two candidates: A and B. Assume that candidate A receives m votes and candidate B receives k votes, where $m > k$. What is the probability that candidate A is always ahead of candidate B in vote counts?

Let the probability that A always leads B in vote counts by a vote of m to k be $P_{m,k}$. Since the total number of votes cast is $n = m + k$ and the difference in votes is $m - k = x$, then from the earlier result we have that

$$P_{m,k} = \frac{x}{n} = \frac{m - k}{m + k} \tag{3.29}$$

3.11.1 The Conditional Probability Method

Let the probability that A is always ahead, given that A received the last vote, be denoted by $P[A|AL]$; let the probability that A is always ahead, given that B received the last vote, be $P[A|BL]$. If $P[AL]$ denotes the probability that A received the last vote, which is $m/(m+k)$ and $P[BL]$ denotes the probability that B received the last vote, which is $k/(m+k)$, then we have that

$$P_{m,k} = P[A \mid AL]P[AL] + P[A \mid BL]P[BL]$$

$$= P[A \mid AL]\left\{\frac{m}{m+k}\right\} + P[A \mid BL]\left\{\frac{k}{m+k}\right\}$$

$$= P_{m-1,k}\left\{\frac{m}{m+k}\right\} + P_{m,k-1}\left\{\frac{k}{m+k}\right\}$$

With the initial condition as $P_{1,0} = 1$, we derive the answer by induction. Define $m + k = L$. Then the equation is true for $L = 1$ because $P_{1,0} = (1-0)/(1+0) = 1$. Similarly, it is true for $L = 2$ because the only valid configuration is $(2, 0)$ and

$$P_{2,0} = P_{1,0}\frac{2}{2+0} = 1 = \frac{2-0}{2+0}$$

Also, when $L = 3$ the valid configurations are $(m, k) = \{(2, 1), (3, 0)\}$. Now, because we want $P_{1,1} = 0$, we obtain

$$P_{2,1} = P_{2,0}\left(\frac{1}{3}\right) = \frac{1}{3} = \frac{2-1}{2+1}$$

$$P_{3,0} = P_{2,0}\left(\frac{3}{3}\right) = 1 = \frac{3-0}{3+0}$$

Assume that the solution holds for L. Let m and k be such that $m \geq k$ and $m+k=L+1$. Then by the induction we have that

$$P_{m,k} = \left\{ \frac{m-1-k}{m-1+k} \right\} \left[\frac{m}{m+k} \right] + \left\{ \frac{m-k+1}{m-1+k} \right\} \left[\frac{k}{m+k} \right] = \frac{m-k}{m+k}$$

Thus, the proposition is also true for $m+k=L+1$ which implies that it is valid for all m and k.

3.12 RETURNS TO THE ORIGIN AND THE ARC-SINE LAW

Consider a symmetric random walk (with $p=1/2$). Recall that $u_{2n}=P[Y_{2n}=0|Y_n=0]$ is the probability that the walker returns to the origin. Let f_{2n} denote the probability of the first return to 0 at time $2n$; that is,

$$f_{2n} = P[Y_{2n} = 0 \mid Y_0 = 0, Y_k \neq 0, k = 1, \ldots, 2n-1]$$

As stated earlier, in order to return to the origin at time $2n$ the walker must take n steps to the right and n steps to the left. Thus,

$$u_{2n} = \binom{2n}{n}\left(\frac{1}{2}\right)^{2n}$$

To compute f_{2n}, we observe that a visit to the origin at time $2n$ may be the first return or the first return occurs at time $2k<2n$ followed by a return to the origin at time $2n$, which is $2n-2k$ time units later. Thus, we have the following relationship:

$$
\begin{aligned}
u_{2n} &= f_{2n}u_0 + f_{2n-2}u_2 + \cdots + f_4 u_{2n-4} + f_2 u_{2n-2} \\
&= \sum_{k=1}^{n} f_{2k} u_{2n-2k}
\end{aligned}
\tag{3.30}
$$

We define the z-transforms

$$F(z) = \sum_{n=1}^{\infty} f_{2n} z^n$$

$$U(z) = \sum_{n=0}^{\infty} u_{2n} z^n$$

Since $u_0=1$ and $F(z)$ is defined for $n \geq 1$, then from Equation 3.30 we obtain

$$U(z) = 1 + U(z)F(z) \Rightarrow F(z) = \frac{U(z)-1}{U(z)} \qquad (3.31)$$

Now, we know that

$$u_{2n} = \binom{2n}{n}\left(\frac{1}{2}\right)^{2n} \Rightarrow U(z) = \sum_{n=0}^{\infty}\binom{2n}{n}\left(\frac{1}{2}\right)^{2n} z^n = \sum_{n=0}^{\infty}\binom{2n}{n}\left(2^{-2}z\right)^n$$

As discussed earlier, it is known that

$$\sum_{n=0}^{\infty}\binom{2n}{n}x^n = \frac{1}{\sqrt{1-4x}}$$

Thus,

$$U(z) = \frac{1}{\sqrt{1-4(z/4)}} = \frac{1}{\sqrt{1-z}}$$

From Equation 3.31 we obtain

$$F(z) = \frac{U(z)-1}{U(z)} = \frac{(1-z)^{-1/2}-1}{(1-z)^{-1/2}} = 1 - (1-z)^{1/2} = 1 - \sqrt{1-z}$$

We recall from the binomial theorem that

$$\sqrt{1-z} = \sum_{n=0}^{\infty}\binom{\frac{1}{2}}{n}(-1)^n z^n = 1 + \sum_{n=1}^{\infty}\binom{\frac{1}{2}}{n}(-1)^n z^n$$

From this we obtain

$$F(z) = -\sum_{n=1}^{\infty}\binom{\frac{1}{2}}{n}(-1)^n z^n$$

An Aside: We transform the fractional binomial component $\binom{\frac{1}{2}}{n}$ into non-fractional component as follows:

$$\binom{\frac{1}{2}}{n} = \frac{\frac{1}{2}\left(\frac{1}{2}-1\right)\left(\frac{1}{2}-2\right)\cdots\left(\frac{1}{2}-(n-1)\right)\left(\frac{1}{2}-n\right)!}{n!\left(\frac{1}{2}-n\right)!}$$

$$= \frac{\frac{1}{2}\left(\frac{1}{2}-1\right)\left(\frac{1}{2}-2\right)\cdots\left(\frac{1}{2}-(n-1)\right)}{n!}$$

$$= \frac{\frac{1}{2}\left(-\frac{1}{2}\right)\left(-\frac{3}{2}\right)\cdots\left(-\frac{2n-3}{2}\right)}{n!}$$

$$= \frac{\frac{1}{2}\left(-\frac{1}{2}\right)^{n-1}(1)(3)(5)\ldots(2n-5)(2n-3)}{n!}$$

$$= \frac{1}{2}\left(-\frac{1}{2}\right)^{n-1}\frac{(1)(3)(5)\ldots(2n-5)(2n-3)}{n!}$$

$$= \frac{1}{2}\left(-\frac{1}{2}\right)^{n-1}\frac{(1)(2)(3)(4)(5)\ldots(2n-5)(2n-4)(2n-3)(2n-2)}{n!(2)(4)(6)\ldots(2n-4)(2n-2)}$$

$$= \frac{1}{2}\left(-\frac{1}{2}\right)^{n-1}\frac{(1)(2)(3)(4)(5)\ldots(2n-5)(2n-4)(2n-3)(2n-2)}{n!2^{n-1}(1)(2)(3)\ldots(n-2)(n-1)}$$

$$= \frac{1}{2}\left(-\frac{1}{2}\right)^{n-1}\frac{(2n-2)!}{n!2^{n-1}(n-1)!} = (-1)^{n-1}\frac{(2n-2)!}{n!2^{2n-1}(n-1)!} \qquad (*)$$

$$= (-1)^{n-1}\frac{2n(2n-1)(2n-2)!n}{n!2^{2n-1}n(n-1)!2n(2n-1)} = (-1)^{n-1}\frac{(2n)!}{n!n!2^{2n}(2n-1)}$$

$$= (-1)^{n-1}\binom{2n}{n}\frac{1}{2^{2n}(2n-1)}$$

Thus,

$$F(z) = -\sum_{n=1}^{\infty}\binom{\frac{1}{2}}{n}(-1)^n z^n = \sum_{n=1}^{\infty}\binom{2n}{n}\frac{1}{2^{2n}(2n-1)}z^n$$

$$f_{2n} = \binom{2n}{n}\frac{1}{2^{2n}(2n-1)} \qquad n = 1, 2, \ldots \qquad (3.32)$$

$$f_{2n} = \frac{u_{2m}}{2n-1} \qquad (3.33)$$

Observe from (*) that

$$\binom{\frac{1}{2}}{n} = (-1)^{n-1}\frac{(2n-2)!}{n!2^{2n-1}(n-1)!} = (-1)^{n-1}\frac{(2n-2)!}{n(n-1)!2^{2n-1}(n-1)!}$$

$$= (-1)^{n-1}\binom{2n-2}{n-1}\left(\frac{1}{2}\right)^{2n-2}\left(\frac{1}{2n}\right) = (-1)^{n-1}\frac{u_{2n-2}}{2n}$$

Thus,

$$F(z) = -\sum_{n=1}^{\infty}\binom{\frac{1}{2}}{n}(-1)^n z^n = \sum_{n=1}^{\infty}\frac{u_{2n-2}}{2n}z^n$$

This means that

$$f_{2n} = \frac{u_{2n-2}}{2n} \tag{3.34}$$

Proposition 3.1: The probability that from time 1 to $2n$ the walker does not return to the origin is u_{2n}; that is,

$$P[Y_1 \neq 0, Y_2 \neq 0, ..., Y_n \neq 0] = u_{2n} \tag{3.35}$$

Let $P[Y_{2n}=2k]=p_{2n,2k}$ denote the probability that the latest return to the origin of a walk of length $2n$ occurs at time $2k$.

Proposition 3.2: $p_{2n,2k} = u_{2k}u_{2n-2k} = \binom{2k}{k}\binom{2n-2k}{n-k}\left(\frac{1}{2}\right)^{2n}$ $\quad k = 0, 1, ..., n$

Proof:

$$P[Y_{2n} = 2k] = P[Y_{2k} = 0, Y_{2k+1} \neq 0, Y_{2k+2} \neq 0, ..., Y_{2n} \neq 0]$$
$$= P[Y_{2k} = 0]P[Y_{2k+1} \neq 0, Y_{2k+2} \neq 0, ..., Y_{2n} \neq 0 \,|\, Y_{2k} = 0]$$
$$= P[Y_{2k} = 0]P[Y_1 \neq 0, Y_2 \neq 0, ..., Y_{2n-2k} \neq 0]$$
$$= P[Y_{2k} = 0]P[Y_{2n-2k} = 0]$$
$$= u_{2k}u_{2n-2k}$$

where the second to the last equality follows from Equation 3.35. This proposition simply states that the walk consists of a loop of length $2k$ followed

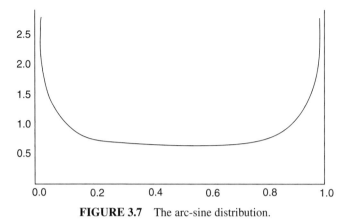

FIGURE 3.7   The arc-sine distribution.

by a path of length $2n-2k$ with no return. Observe that $p_{2n,\,2k}$ is symmetrical in the sense that

$$p_{2n,\,2k} = p_{2n-2k,\,2n}, \quad k = 0, 1, \ldots, n$$

That is, the value of $p_{2n,2k}$ is the same for k and $n-k$. This implies that if the walker is on the positive side at half time, then with probability one half he will return to the positive side for the remainder of the walk.

We can use the Stirling's formula to obtain the limiting value of $p_{2n,\,2k}$ as $n \to \infty$ as follows. Recall that $n! \sim (n/e)^n \sqrt{2\pi n}$ for $n \to \infty$. Thus,

$$
\begin{aligned}
p_{2n,2k} &= \binom{2k}{k}\binom{2n-2k}{n-k}\left(\frac{1}{2}\right)^{2n} = \frac{(2k)!(2n-2k)!}{(k!)^2\left((n-k)!\right)^2}\left(\frac{1}{2}\right)^{2n} \\[2mm]
&\sim \frac{(2k/e)^{2k}\left\{\sqrt{4\pi k}\right\}\{(2n-2k)/e\}^{2n-2k}\left\{\sqrt{4\pi(n-k)}\right\}}{(k/e)^{2k}\left\{\sqrt{2\pi k}\right\}^2\{(n-k)/e\}^{2n-2k}\left\{\sqrt{2\pi(n-k)}\right\}^2}\left(\frac{1}{2}\right)^{2n} \quad (3.36)\\[2mm]
&= \frac{1}{\pi\sqrt{k(n-k)}} = \frac{1}{n\pi\sqrt{\frac{k}{n}(1-\frac{k}{n})}}
\end{aligned}
$$

This is the so-called arc-sine law. If we define $l=k/n$, the last function becomes

$$f(l) = \frac{1}{n\pi\sqrt{l(1-l)}}$$

which is a U-shaped function that is flat in the middle of $[0,1]$ and goes to infinity at the endpoints, as shown in Figure 3.7.

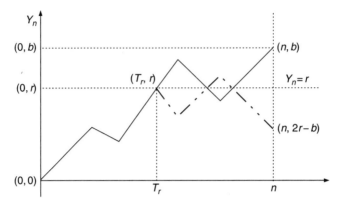

FIGURE 3.8   Illustration of reflection principle for maximum of random walk.

3.13   MAXIMUM OF A RANDOM WALK

Let M_n denote the maximum of a symmetric random walk up to time n; that is,

$$M_n = \max\{Y_k : k = 1, 2, \ldots, n\}$$

Theorem 3.2: For a symmetric random walk and any $r \geq 1$,

$$P[M_n \geq r, Y_n = b] = \begin{cases} P[Y_n = b] & b \geq r \\ P[Y_n = 2r \ b] & b < r \end{cases}$$

Proof: Let T_r be the time at which the path touches the line $Y=r$ for the first time. For every path with $Y_n \geq r$, there exists, according to the reflection principle, another path obtained by reflection through the line $Y_n = r$ from T_r onward up to n, as shown illustrated in Figure 3.8.

From the figure we observe that the case when $b \geq r$ is obvious. For the case when $b < r$, we proceed as follows. Let $N_{0,n}^r(0, b)$ denote the number of paths from $(0, 0)$ to (n, b) that includes some points on the line $Y=r$; that is, some point (i, r), $0 < i < n$. For such a path we reflect the segment with $T_r \leq x \leq n$ on the line $Y_n = r$ to obtain a path from $(0, 0)$ to $(n, 2r-b)$. Let $N_{0,n}(0, 2r-b)$ be the number of paths from $(0, 0)$ to $(n, 2r-b)$. Then according to the reflection principle, $N_{0,n}^r(0, b) = N_{0,n}(0, 2r-b)$. Thus,

$$P[M_n \geq r, Y_n = b] = N_{0,n}^r(0, b)\left(\frac{1}{2}\right)^n$$

$$= N_{0,n}(0, 2r-b)\left(\frac{1}{2}\right)^n$$

$$= P[Y_n = 2r-b]$$

This completes the proof. From this we obtain

$$P[M_k \geq r] = \sum_{b=-\infty}^{\infty} P[M_k \geq r, Y_k = b]$$

$$= \sum_{b=r}^{\infty} P[M_k \geq r, Y_k = b] + \sum_{b=-\infty}^{r-1} P[M_k \geq r, Y_k = b]$$

$$= \sum_{b=r}^{\infty} P[Y_k = b] + \sum_{b=-\infty}^{r-1} P[Y_k = 2r - b]$$

$$= P[Y_k \geq r] + \sum_{m=r+1}^{\infty} P[Y_k = m] \qquad (3.37)$$

$$= P[Y_k = r] + \sum_{m=r+1}^{\infty} P[Y_k = m] + \sum_{m=r+1}^{\infty} P[Y_k = m]$$

$$= P[Y_k = r] + 2 \sum_{m=r+1}^{\infty} P[Y_k = m]$$

$$= P[Y_k = r] + 2P[Y_k > r]$$

3.14 TWO SYMMETRIC RANDOM WALKERS

Consider a situation where two walkers start out together at the origin at time 0 executing independent symmetric random walks. We are interested in the probability that they meet again after each has taken N steps.

To solve this problem, we realize that for the walkers to meet again, each of them must have taken the same number of steps to the right out of the N steps. This is an example of the *exchangeability* property, which in this case essentially means that the exact order of their right and left steps is unimportant; what matters is that the number of steps to the right out of the N steps is the same for both walkers. Let the number of steps to the right for each walker be r, which means that the number of steps to the left is $N-r$. Then there are $\binom{N}{r}$ ways that the r steps to the right can occur among the N total number of steps, each of which occurs with probability $\left(\frac{1}{2}\right)^N$. Since the two walkers operate independently, the probability that they meet again after N steps is given by

$$P_N = \sum_{r=0}^{N} \left\{ \binom{N}{r} \frac{1}{2^N} \right\}^2 = \frac{1}{2^{2N}} \sum_{r=0}^{N} \binom{N}{r}^2 = \frac{1}{2^{2N}} \binom{2N}{N} = \frac{(2N)!}{\left(2^N N!\right)^2}$$

3.15 RANDOM WALK ON A GRAPH

A graph $G=(V, E)$ is a pair of sets V (or $V(G)$) and E (or $E(G)$) called vertices (or nodes) and edges (or arcs), respectively, where the edges join different pairs of

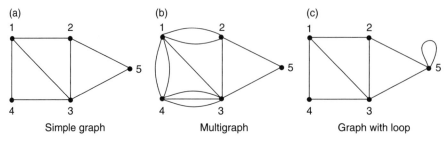

(a) 1 2 5 4 3
Simple graph

(b) 1 2 5 4 3
Multigraph

(c) 1 2 5 4 3
Graph with loop

FIGURE 3.9 Examples of simple graph, multigraph and graph with loop.

vertices. The vertices are represented by points, and the edges are represented by lines joining the vertices. A graph is a mathematical concept that captures the notion of connection. A graph is defined to be a *simple graph* if there is at most one edge connecting any pair of vertices and an edge does not loop to connect a vertex to itself. When multiple edges are allowed between any pair of vertices, the graph is called a *multigraph*. Examples of a simple graph, a multigraph, and a graph with loop are shown in Figure 3.9.

Two vertices are said to be *adjacent* if they are joined by an edge. For example, in Figure 3.9, vertices 1 and 2 are adjacent. An edge e that connects vertices a and b is denoted by (a, b). Such an edge is said to be *incident* with vertices a and b; the vertices a and b are called the *ends* or *endpoints* of e. If the edge $e=(a, b)$ exists, we sometimes call vertex b a *neighbor* of vertex a. The set of neighbors of vertex a is usually denoted by $\Gamma(a)$.

The *degree* (or *valency*) of a vertex x, which is denoted by $d(x)$, is the number of edges that are incident with x. For example, in Figure 3.9a, $d(3)=4$ and $d(4)=2$. If a node x has $d(x)=0$, then x is said to be *isolated*. It can be shown that

$$\sum_{x \in V(G)} d(x) = 2|E(G)| \tag{3.38}$$

where $|E(G)|$ is the number of edges in the graph.

A *subgraph* of G is a graph H such that $V(H) \subseteq V(G)$ and $E(H) \subseteq E(G)$, and the endpoints of an edge $e \in E(H)$ are the same as its endpoints in G. A *complete graph* K_n on n vertices is the simple graph that has all $\binom{n}{2}$ possible edges.

A *walk* in a graph is an alternating sequence $x_0, e_1, x_1, e_2, \ldots, x_{k-1}, e_k, x_k$ of vertices x_i, which are not necessarily distinct, and edges e_i such that the endpoints of e_i are x_{i-1} and x_i, $i=1, \ldots, k$. A *path* is a walk in which the vertices are distinct. For example, in Figure 3.9a, the path $\{1,3,5\}$ connects vertices 1 and 5. When a path can be found between every pair of distinct vertices, we say that the graph is a *connected graph*. A graph that is not connected can be decomposed into two or more connected subgraphs, each pair of which has no node in common. That is, a disconnected graph is the union of two or more disjoint connected subgraphs.

Another way to describe a graph is in terms of the *adjacency matrix* $A(x, y)$, which has a value 1 in its cell if x and y are neighbors and zero otherwise, for all $x, y \in V$. Then the degree of vertex x is given by

$$d(x) = \sum_y A(x, y) \qquad (3.39)$$

3.15.1 Proximity Measures

The proximity measures for connected graphs include the following:

- The *hitting time* from node v_i to node v_j is denoted by $H(v_i, v_j)$ and defined as the expected number of steps required to reach v_j for the first time from v_i. The hitting time is not symmetric because generally $H(v_i, v_j) \neq H(v_j, v_i)$.
- The *cover time* $C(v_i)$ from node v_i is the expected number of steps required to visit all the nodes starting from v_i. The cover time for a graph is the maximum $C(v_i)$ over all nodes v_i and denoted by $C(G)$.
- The *commute time* $C(v_i, v_j)$ between node v_i and node v_j is the expected number of steps that it takes to go from v_i to v_j and back to v_i. That is,

$$C(v_i, v_j) = H(v_i, v_j) + H(v_j, v_i) \qquad (3.40)$$

The commute time is symmetric in the sense that $C(v_i, v_j) = C(v_j, v_i)$.

A bound for $C(G)$ was obtained by Kahn (1989) as $C(G) \leq 4n^2 d_{ave}/d_{min}$, where n is the number of nodes in the graph, d_{ave} is the average degree of the graph, and d_{min} is the minimum degree of the graph. In Bollobas (1998) it is shown that in a connected graph with m edges, the mean *return time* to a vertex v, which is denoted by $H(v, v)$, is given by

$$H(v, v) = \frac{2m}{d(v)} \qquad (3.41)$$

3.15.2 Directed Graphs

Graphs are often used to model relationships. When there is a special association in these relationships, the *undirected graphs* we have described so far do not convey this information; a *directed graph* is required. A directed graph (or *digraph*) is a graph in which an edge consists of an ordered vertex pair, giving it a direction from one vertex to the other. Generally in a digraph, the edge (a, b) has a direction from vertex a to vertex b, which is indicated by an arrow in the direction from a to b. Figure 3.10 illustrates a simple digraph. When the directions are ignored, we obtain the underlying undirected graph shown in Figure 3.9**a**.

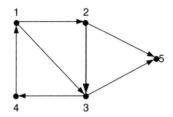

FIGURE 3.10   Example of a digraph.

3.15.3   Random Walk on an Undirected Graph

Let $G=(V, E)$ be a connected undirected graph with n vertices and m edges. A random walk on G can be described as follows. We start at vertex v_0 and arrive at vertex v_i in the kth step. We move to vertex v_j, which is one of the neighbors of vertex v_i, with probability $1/d(v_i)$. The sequence of random vertices $\{v_t, t=0, 1, \ldots\}$ is a Markov chain with transition probabilities p_{ij} given by

$$p_{ij} = \begin{cases} 1/d(i) & i, j \in V \\ 0 & \text{otherwise} \end{cases} \tag{3.42}$$

Let $P=[p_{ij}]_{i, j \in V}$ be the state transition probability matrix. Then for the simple graph in Figure 3.9a, we have that

$$P = \begin{bmatrix} 0 & 1/3 & 1/3 & 1/3 & 0 \\ 1/3 & 0 & 1/3 & 0 & 1/3 \\ 1/4 & 1/4 & 0 & 1/4 & 1/4 \\ 1/2 & 0 & 1/2 & 0 & 0 \\ 0 & 1/2 & 1/2 & 0 & 0 \end{bmatrix}$$

which corresponds to the state transition diagram shown in Figure 3.11.

Random walk on a graph is used as a search technique in which a search proceeds from a start node by randomly selecting one of its neighbors, say k. At k the search randomly selects one of its neighbors, making an effort not to reselect the node from where it reached k, and so on. If the goal is to reach a particular destination node, the search terminates when this destination is reached.

We can construct the Markov chain of the multigraph in a similar manner. In this case,

$$p_{ij} = \begin{cases} n_{ij}/d(i) & i, j \in V \\ 0 & \text{otherwise} \end{cases} \tag{3.43}$$

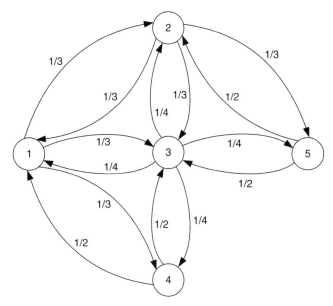

FIGURE 3.11   State-transition diagram of graph in Figure 3.9**a**.

where n_{ij} is the number of edges between nodes i and j. For example, in the multi-graph of Figure 3.9**b**, we have that

$$P = \begin{bmatrix} 0 & 2/5 & 1/5 & 2/5 & 0 \\ 1/2 & 0 & 1/4 & 0 & 1/4 \\ 1/6 & 1/6 & 0 & 1/2 & 1/6 \\ 2/5 & 0 & 3/5 & 0 & 0 \\ 0 & 1/2 & 1/2 & 0 & 0 \end{bmatrix}$$

The Markov chain of the multigraph is shown in Figure 3.12.

The Markov chain associated with a random walk on a graph is irreducible if and only if the graph is connected. Let m denote the number of edges in an undirected connected graph $G=(V, E)$, and let $\{\pi_k, k \in V\}$ be the stationary distribution of the Markov chain associated with the graph. The stationary distribution of the Markov chain associated with $G=(V, E)$ is given by the following theorem:

Theorem 3.3: The stationary distribution of the Markov chain associated with the connected graph $G=(V, E)$ is given by $\pi_i = d(i)/2m$, $i=1, \ldots, n$; where m is the number of edges in the graph, as defined earlier.

Proof: The proof consists in our showing that the distribution $\pi=(\pi_1, \ldots, \pi_n)$ satisfies the equation $\pi P = \pi$. We prove the theorem with a multigraph, which is more general than the simple graph. Thus, we have that with respect to node j,

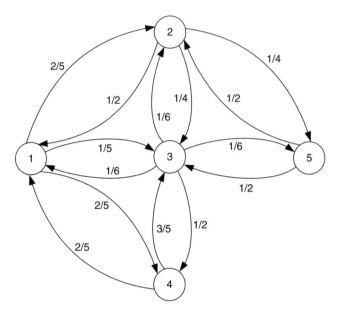

FIGURE 3.12 State-transition diagram of multigraph in Figure 3.9b.

$$(\pi P)_j = \sum_i \pi_i p_{ij} = \sum_i \left\{ \frac{d(i)}{2m} \times \frac{n_{ij}}{d(i)} \right\} = \frac{1}{2m} \sum_i n_{ij} = \frac{d(j)}{2m} = \pi_j$$

This implies that by definition π is the stationary distribution of the unique Markov chain defined by P. This completes the proof.

Note that for the simple graph we have that $n_{ij}=1$, and the same result holds.

Example 3.1: Consider the simple graph of Figure 3.9a. We have that $m=7$, which means that the stationary distribution is given by

$$\pi = \left(\frac{3}{14}, \frac{3}{14}, \frac{4}{14}, \frac{2}{14}, \frac{2}{14} \right)$$

Similarly, for the multigraph of Figure 3.9b, the number of edges is $m=11$. Thus, the stationary distribution of the Markov chain in Figure 3.11 is given by

$$\pi = \left(\frac{5}{22}, \frac{4}{22}, \frac{6}{22}, \frac{5}{22}, \frac{2}{22} \right)$$

Note that a loop is considered to contribute twice to the degree of a node. Thus, in the case of the graph with loop shown in Figure 3.9c, $m=8$, and because $d(5)=4$, we obtain the stationary distribution as follows:

$$\pi = \left(\frac{3}{16}, \frac{3}{16}, \frac{4}{16}, \frac{2}{16}, \frac{4}{16} \right)$$

Recall that the mean return time to a node v in a connected graph with m edges is given by $H(v, v) = 2m/d(v)$. From the results on the stationary distributions we may then write

$$H(v, v) = \frac{1}{\pi_v}$$

Example 3.2: Consider a random walk on a two-dimensional lattice consisting of the 4×4 checkerboard shown in Figure 3.13.

The number of edges is $m = 24$, and the degrees of the nodes are as follows:

$$d(1) = d(4) = d(13) = d(16) = 2$$
$$d(2) = d(3) = d(5) = d(8) = d(9) = d(12) = d(14) = d(15) = 3$$
$$d(6) = d(7) = d(10) = d(11) = 4$$

Thus, the stationary probabilities are

$$\pi_1 = \pi_4 = \pi_{13} = \pi_{16} = \frac{1}{24}$$

$$\pi_2 = \pi_3 = \pi_5 = \pi_8 = \pi_9 = \pi_{12} = \pi_{14} = \pi_{15} = \frac{1}{16}$$

$$\pi_6 = \pi_7 = \pi_{10} = \pi_{11} = \frac{1}{12}$$

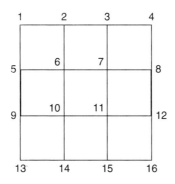

FIGURE 3.13   Checkerboard.

3.15.4 Random Walk on a Weighted Graph

A more general random walk on a graph is that performed on a weighted graph. Specifically, we consider a connected graph $G=(V, E)$ with positive weight w_e assigned to edge $e \in E$. The weighted random walk is a random walk where the transition probabilities are proportional to the weights of the edges; that is,

$$p_{ij} = \frac{w_{(i, j)}}{\sum_{k \sim i} w_{(i, k)}} \equiv \frac{w_{ij}}{\sum_{k \sim i} w_{ik}} \qquad (3.44)$$

If all the weights are 1, we obtain a simple random walk. Let the total weight of the edges emanating from node i be w_i, which is given by

$$w_i = \sum_{k \sim i} w_{ik}$$

Then the sum of the weights of all edges is

$$w = \sum_{i, k | k > i} w_{ik}$$

where the inequality in the summation is used to avoid double counting. It is easy to show that the stationary distribution is given by

$$\pi_i = \frac{w_i}{2w}$$

The proof consists as usual in verifying that the preceding distribution satisfies the relationship $\pi P = \pi$, which can be seen as follows:

$$\sum_i \pi_i p_{ij} = \sum_i \left\{ \frac{w_i}{2w} \times \frac{w_{ij}}{w_i} \right\} = \sum_i \frac{w_{ij}}{2w} = \frac{1}{2w} \sum_i w_{ij}$$

$$= \frac{w_j}{2w} = \pi_j$$

Thus, the stationary probability of state i is proportional to the weight of the edges emanating from node i.

3.16 RANDOM WALKS AND ELECTRIC NETWORKS

In this section, we discuss the relationship between random walks and electric networks. An electric network is a connected undirected graph in which each edge (x, y) has a weight called resistance, $r_{xy} > 0$. It is a symmetric system in the sense that $r_{xy} = r_{yx}$.

The reciprocal of resistance is called *conductance* c_{xy}. That is, $c_{xy} = 1/r_{xy}$. A random walk on the underlying graph of the electric network is defined by assigning a probability to each edge (x, y) that is incident to node x. Thus, for edge (x, y) we have that

$$p_{xy} = \frac{c_{xy}}{c_x} \tag{3.45}$$

where the normalizing constant c_x is given by

$$c_x = \sum_{y \in \Gamma(x)} c_{xy}$$

While $c_{xy} = c_{yx}$, p_{xy} is not necessarily equal to p_{yx} because of the normalization performed at each node. Let $P = [p_{xy}]$ be the transition probability matrix whose elements are the p_{xy}. If the electric network is connected, then the Markov chain with transition probability matrix P is connected and has a stationary probability distribution $\pi = (\pi_1, \pi_2, \ldots)$, where $\pi_x = c_x/c$; c is given by

$$c = \sum_x c_x \tag{3.46}$$

Now,

$$(\pi P)_x = \sum_y \pi_y p_{yx} = \sum_y \frac{c_y}{c} \frac{c_{yx}}{c_y} = \sum_y \frac{c_{yx}}{c} = \frac{c_x}{c} = \pi_x$$

This implies that $\pi = \pi P$. In the case where each edge has a resistance of 1 unit, we have that

$$c_x = \sum_y c_{xy} = d_x$$

$$c = \sum_x c_x = 2m$$

where d_x is the degree is node x and m is the number of edges.

Consider two special nodes in the network labeled a and b, and let the voltage v_b at node b be set to 0 while the voltage at node a, v_a be nonzero; we set $v_a = 1$. Fixing these two voltages induces voltages at other nodes in the network as well as currents flowing along the edges. Consider an arbitrary node x in the network that is adjacent to node y. According to Ohm's law, the current flowing from node x to node y along the edge (x, y) is given by

$$i_{xy} = \frac{v_x - v_y}{r_{xy}} = (v_x - v_y)c_{xy} \tag{3.47}$$

According to the Kirchhoff's current law, the total current flowing into any node is 0; that is,

$$\sum_y i_{xy} = \sum_y (v_x - v_y) c_{xy} = 0 \Rightarrow v_x \sum_y c_{xy} = \sum_y v_y c_{xy}$$

But $\sum_y c_{xy} = c_x$, which means that we have that

$$v_x c_x = \sum_y v_y c_{xy} \Rightarrow v_x = \sum_y v_y \frac{c_{xy}}{c_x} = \sum_y v_y p_{xy}$$

Thus we obtain

$$v_x = \sum_y p_{xy} v_y \tag{3.48}$$

This shows that the voltage at node x is a weighted average of the voltages at the adjacent nodes.

3.16.1 Harmonic Functions

Let P be the transition probability matrix of an irreducible Markov chain with state space Ω. A function $h : \Omega \to \Re$ is defined to be harmonic at x if

$$h(x) = \sum_{y \in \Omega} P(x, y) h(y) \tag{3.49}$$

A harmonic function on an undirected graph partitions the nodes into boundary nodes and interior nodes. For boundary nodes, the value of the harmonic function is fixed. For an interior node x, the value $h(x)$ is the weighted average of h at the neighbors of x. A harmonic function $h(x)$ defined on Ω takes on its maximum and its minimum values on the boundary.

Thus, if we define S to be the set of interior nodes and B the set of boundary nodes such that $S \cup B = \Omega$, then for the electric network we discussed earlier, $B = \{a, b\}$. Since the maximum voltage is $v_a = 1$, the minimum is $v_b = 0$, and $v_x = \sum_y p_{xy} v_y$ at every interior node x, it follows that v_x is a harmonic function.

3.16.2 Effective Resistance and Escape Probability

Consider the setup we defined earlier in which we set $v_a = 1$ and $v_b = 0$. Let i_a be the current flowing into the network at node a and out at node b. The effective resistance r_{eff} between nodes a and b is the resistance between a and b offered by the whole network. That is,

$$r_{\mathrm{eff}} = \frac{v_a - v_b}{i_a} = \frac{v_a}{i_a} \tag{3.50}$$

This means that the network can be considered as a single resistor between a and b, and the resistance of this resistor is the equivalent resistance. It can also be defined as the potential difference between a and b when a current of 1 ampere is injected into node a and removed at b. The *effective conductance* $c_{eff} = 1/r_{eff}$.

We define the *escape probability* p_{esc} as the probability that a random walk that starts at node a reaches node b before returning to node a. The current at node a is given by the Kirchhoff's current law:

$$i_a = \sum_{y \in \Gamma(a)} (v_a - v_y) c_{ay}$$

Because $v_a = 1$, we have that

$$i_a = \sum_{y \in \Gamma(a)} (1 - v_y) c_{ay} = \sum_{y \in \Gamma(a)} (1 - v_y) \frac{c_{ay}}{c_a} c_a = c_a \left\{ \sum_{y \in \Gamma(a)} \frac{c_{ay}}{c_a} - \sum_{y \in \Gamma(a)} \frac{c_{ay}}{c_a} v_y \right\}$$

$$= c_a \left\{ 1 - \sum_{y \in \Gamma(a)} p_{ay} v_y \right\}$$

For each node $y \in \Gamma(a)$, p_{ay} is the probability that a walk that starts at node a goes to node y next. The parameter v_y is the probability that a walk that starts at node y reaches node a before reaching node b. This implies that $\sum_{y \in \Gamma(a)} p_{ay} v_y$ is the probability that a walk that starts at node a returns to node a before reaching node b. Thus, $1 - \sum_{y \in \Gamma(a)} p_{ay} v_y$ is the probability that a walk that starts at node a reaches node b before returning to node a. For this reason we may write $i_a = c_a p_{esc}$. Because $i_a = (v_a - v_b)/r_{eff} = c_{eff}$ since $v_a = 1$ and $v_b = 0$, we have that

$$i_a = c_a p_{esc} = c_{eff} \Rightarrow p_{esc} = \frac{c_{eff}}{c_a}$$

The effective resistance between two nodes is a useful measure of how close the two nodes are; that is, it is a distance measure. A distance $\delta(a, b)$ is any function on pairs of nodes a and b such that:

a. $\delta(a, a) = 0$ for every node a
b. $\delta(a, b) \geq 0$ for any two nodes a and b
c. $\delta(a, b) = \delta(b, a)$; which means that it is a symmetric function
d. $\delta(a, c) \leq \delta(a, b) + \delta(b, c)$ which is the so-called triangle inequality.

To compute the effective resistance we use these two rules:
1. If two or more resistors are connected in series, they may be replaced by a single resistor whose resistance is the sum of their resistances. That is,

$$r_{eff} = r_1 + r_2 + \cdots + r_n$$

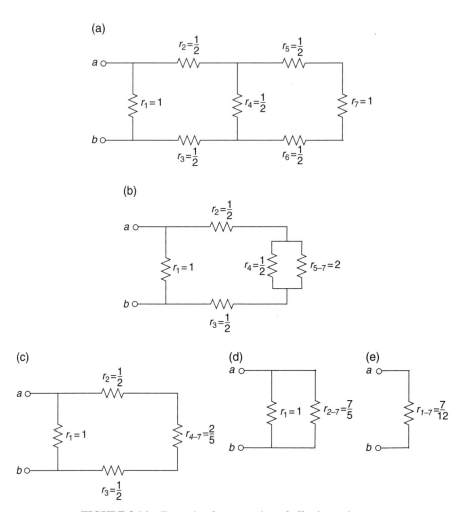

FIGURE 3.14   Example of computation of effective resistance.

2. If two or more resistors are in parallel, they may be replaced by a single resistor whose conductance is the sum of their conductances. That is,

$$c_{\text{eff}} = \frac{1}{r_{\text{eff}}} = c_1 + c_2 + \cdots + c_n = \frac{1}{r_1} + \frac{1}{r_2} + \cdots + \frac{1}{r_n} \Rightarrow r_{\text{eff}} = \frac{1}{\dfrac{1}{r_1} + \dfrac{1}{r_2} + \cdots + \dfrac{1}{r_n}}$$

Example 3.3: We compute the effective resistance of the electric network shown in Figure 3.14. Figure 3.14a is the original network. Figure 3.14b is obtained from the observation that r_5, r_6, and r_7 are in series in Figure 3.14a. Thus, their equivalent resistance is $r_{5-7} = r_5 + r_6 + r_7 = 2$, which is in parallel with r_4. The equivalent resistance of r_4 and r_{5-7} is $r_{4-7}=((1/r_4)+(1/r_{5-7}))^{-1}=2/5$, as shown in Figure 3.14c. In the same way, from Figure 3.14c we find that r_2, r_3, and r_{4-7} are in series. Thus, their equivalent resistance is $r_{2-7}=r_2+r_3+r_{4-7}=7/5$, as shown in Figure 3.14d. Finally, from Figure 3.14d we observe that r_1 and r_{2-7} are in parallel, and their equivalent resistance $r_{1-7}=((1/r_1)+(1/r_{2-7}))^{-1}=7/12$, which is the effective resistance between nodes a and b as shown in Figure 3.14e. That is, $r_{eff}=7/12$; alternatively, $c_{eff}=1/r_{eff}=12/7$. Since $c_a=2+1=3$, the escape probability is

$$p_{esc}=\frac{c_{eff}}{c_a}=\frac{12}{7}\times\frac{1}{3}=\frac{4}{7}$$

3.17 CORRELATED RANDOM WALK

In the classical random walk described earlier, there is a constant probability p of incrementing the current state by 1 and, therefore, a constant probability $q=1-p$ of decrementing the current state by 1. This is due to the fact that the outcomes of the experiments are independent. A more general case is when there is a correlation between results of successive trials; that is, the outcome of the current trial depends on the outcome of the previous trial. Such random walks are called *correlated random walks* (CRWs) or *persistent random walks* and have been studied by many authors including Goldstein (1951), Gillis (1955), Mohan (1955), Seth (1963), Renshaw and Henderson (1981), Kehr and Argyrakis (1986), Lal and Bhat (1989), Argyrakis and Kehr (1992), Hanneken and Franceschetti (1998), and Bohm (2000).

To illustrate this process, we consider an experiment with two possible outcomes: a win or a loss. The probability of the first outcome is assumed to be fixed, and thereafter each subsequent outcome is governed by the following rule. Given that the current trial results in a win, the probability that the next trial will result in a win is p_1, and the probability that it will result in a loss is $q_1=1-p_1$. Similarly, given the current trial results in a loss, the probability that the next trial will result in a loss is p_0, and the probability that it will result in a win is $q_0=1-p_0$. Thus, the conditional probabilities are as follows:

$$P[\text{win}|\text{win}] = p_1$$
$$P[\text{loss}|\text{win}] = q_1 = 1-p_1$$
$$P[\text{loss}|\text{loss}] = p_0$$
$$P[\text{win}|\text{loss}] = q_0 = 1-p_0$$

The CRW can be modeled as a bivariate Markov chain with the location of the walker and the result of the previous walk as the two variables. Specifically, we represent a state by (k, l), where k is the location of the walker, where a move to the right indicates a success and a move to the left indicates a failure; and l is the result index that is defined by

(a)

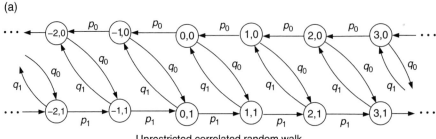

Unrestricted correlated random walk

(b)

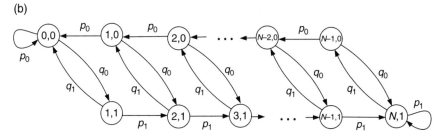

Correlated random walk with reflecting barriers

(c)

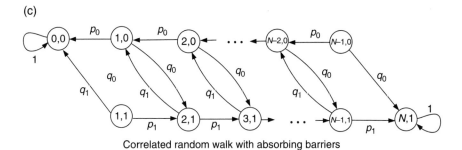

Correlated random walk with absorbing barriers

FIGURE 3.15   State-transition diagrams for correlated random walks.

$$l = \begin{cases} 1 & \text{if the previous trial resulted in a success} \\ 0 & \text{if the previous trial resulted in a failure} \end{cases}$$

The range of k depends on the type of CRW. When the walk is unrestricted, then $k = \ldots, -2, -1, 0, 1, 2, \ldots$. Similarly, when barriers exist, we have that $k = 0, 1, 2, \ldots, N$, where N can be infinite. Thus, the state-transition diagrams in Figure 3.15 show the cases of unrestricted CRW, CRW with reflecting barriers, and CRW with absorbing barriers.

As can be seen from Figure 3.15, the CRW is a quasi-birth-and-death process. Let $\pi_{k,0}(n)$ denote the probability of being in state $(k, 0)$ after n trials, and let $\pi_{k,1}(n)$ denote

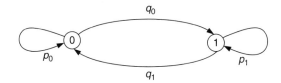

FIGURE 3.16   State-transition diagrams for result sequence.

the probability of being in state $(k, 1)$ after n trials. Then we have the following difference equations:

$$\pi_{k,0}(n+1) = p_0\pi_{k+1,0}(n) + q_1\pi_{k+1,1}(n) \tag{3.52}$$

$$\pi_{k,1}(n+1) = q_0\pi_{k-1,0}(n) + p_1\pi_{k-1,1}(n) \tag{3.53}$$

Let the characteristic function of $\pi_{k,l}(n)$ be defined by

$$\Pi_l(w, n) = \sum_{k=-\infty}^{\infty} w^k \pi_{k,l}(n) \quad l = 0, 1; \quad n = 1, 2, \dots$$

Taking the characteristic functions of both sides of the difference equations, we obtain

$$\Pi_0(w, n+1) = \frac{p_0}{w}\Pi_0(w, n) + \frac{q_1}{w}\Pi_1(w, n)$$
$$\Pi_1(w, n+1) = wq_0\Pi_0(w, n) + wp_1\Pi_1(w, n)$$

which can be arranged in the following form:

$$\begin{bmatrix} \Pi_0(w, n+1) \\ \Pi_1(w, n+1) \end{bmatrix} = \begin{bmatrix} w^{-1}p_0 & w^{-1}q_1 \\ wq_0 & wp_1 \end{bmatrix} \begin{bmatrix} \Pi_0(w, n) \\ \Pi_1(w, n) \end{bmatrix}$$

The initial states $\pi_{k,0}(0)$ and $\pi_{k,1}(0)$ can be obtained by noting that the sequence of result indices $\{l\}$ of the state (k, l) constitutes a two-state Markov chain shown in Figure 3.16.

Thus, if we let $\pi_0(0)$ denote the probability that the process initially started with the first trial being a failure and $\pi_1(0)$ the probability that the first trial was a success, then we have that

$$\pi_0(0) = \frac{q_1}{q_0 + q_1}$$

$$\pi_1(0) = \frac{q_0}{q_0 + q_1}$$

The exact solution of the transform equation of the correlated walk after n trials is of the form

$$\Pi(w, n) = \begin{bmatrix} \Pi_0(w, n) \\ \Pi_1(w, n) \end{bmatrix} = \begin{bmatrix} w^{-1}p_0 & w^{-1}q_1 \\ wq_0 & wp_1 \end{bmatrix}^n \begin{bmatrix} \pi_0(0) \\ \pi_1(0) \end{bmatrix}$$

Let A be the coefficient matrix; that is,

$$A = \begin{bmatrix} w^{-1}p_0 & w^{-1}q_1 \\ wq_0 & wp_1 \end{bmatrix}$$

Then we have that

$$\Pi(w, n) = A^n \pi(0) \tag{3.54}$$

We solve the equation via the diagonalization method, which requires our obtaining the eigenvalues of A. The eigenvalues λ of A satisfy the equation

$$|\lambda I - A| = 0$$

which gives

$$\lambda^2 - \left(wp_1 + w^{-1}p_0\right)\lambda + p_0p_1 - q_0q_1 = 0$$

from which we obtain

$$\lambda = \frac{\left(wp_1 + w^{-1}p_0\right) \pm \sqrt{\left(wp_1 - w^{-1}p_0\right)^2 + 4q_0q_1}}{2}$$

Thus, the eigenvalues are

$$\lambda_1 = \frac{(wp_1 + w^{-1}p_0) + \sqrt{\left(wp_1 - w^{-1}p_0\right)^2 + 4q_0q_1}}{2}$$

$$\lambda_2 = \frac{(wp_1 + w^{-1}p_0) - \sqrt{\left(wp_1 - w^{-1}p_0\right)^2 + 4q_0q_1}}{2}$$

Let the matrix X be defined by $X=(X_1, X_2)$, where $X_i=(x_{1i}, x_{2i})^T$, $i=1,2$. The eigenvectors belonging to λ_1 are obtained from

$$(\lambda_1 I - A)X_1 = 0$$

That is,

$$\left(\lambda_1 - w^{-1}p_0\right)x_{11} - w^{-1}q_1x_{21} = 0$$
$$-wq_0x_{11} + \left(\lambda_1 - wp_1\right)x_{211} = 0$$

If we set $x_{11}=1$ in the first equation, we obtain $x_{21} = \left(\lambda_1 - w^{-1}p_0\right)/w^{-1}q_1 = \left(\lambda_1 w - p_0\right)/q_1$. Note that $(\lambda_1 - w^{-1}p_0)/w^{-1}q_1 = q_0/(\lambda_1 - wp_1)$ because this relationship yields the equation

$$\lambda_1^2 - \left(wp_1 + w^{-1}p_0\right)\lambda_1 + p_0p_1 - q_0q_1 = 0$$

that we know to be true. Thus,

$$X_1 = \left(x_{11}, \ x_{21}\right)^T = \left[1, \frac{\left(\lambda_1 w - p_0\right)}{q_1}\right]^T$$

is an eigenvector. Similarly, the eigenvectors belonging to λ_2 are obtained from

$$(\lambda_2 I - A)X_2 = 0$$

which is similar to the equation for the eigenvectors belonging to λ_1. Thus, another eigenvector is

$$X_2 = \left(x_{12}, \ x_{22}\right)^T = \left[1, \frac{\left(\lambda_2 w - p_0\right)}{q_1}\right]^T$$

Then using the diagonalization method we obtain

$$\prod(w, \ n) = A^n \pi(0) = S\Lambda^n S^{-1}\pi(0) \tag{3.55}$$

where $A=S\Lambda S^{-1}$, S is the matrix whose columns are the eigenvectors and Λ is the diagonal matrix of the corresponding eigenvalues. That is,

$$S = \begin{bmatrix} 1 & 1 \\ (\lambda_1 w - p_0)/q_1 & (\lambda_2 w - p_0)/q_1 \end{bmatrix}$$

$$S^{-1} = \frac{q_1}{w(\lambda_2 - \lambda_1)}\begin{bmatrix} (\lambda_2 w - p_0)/q_1 & -1 \\ -(\lambda_1 w - p_0)/q_1 & 1 \end{bmatrix}$$

$$\Lambda = \begin{bmatrix} \lambda_1 & 0 \\ 0 & \lambda_2 \end{bmatrix} \Rightarrow \Lambda^n = \begin{bmatrix} \lambda_1^n & 0 \\ 0 & \lambda_2^n \end{bmatrix}$$

From this we can obtain the $\pi_{k,1}(n)$, which are the inverse transforms of the components of $\prod(w, n)$.

CRW is popularly used in ecology to model animal and insect movement. For example, it is used in Kareiva and Shigesada (1983) to analyze insect movement. Similarly, it has been used in Byers (2001) to model animal dispersal and in Jonsen (2005) to model animal movement. CRW has also been used to model mobility in mobile ad hoc networks in Bandyopadhyay (2006).

3.18   CONTINUOUS-TIME RANDOM WALK

In the random walk models described earlier, a walker takes steps in a periodic manner, such as every second or minute or hour, or any other equal time interval. More importantly, a classical random walk is a Bernoulli process that allows only two possible events that have values of ± 1. A more general case is when the time between steps is a random variable and the step size is a random variable. In this case we obtain a continuous-time random walk (CTRW), which was introduced by Montroll and Weiss (1965). CTRW has been applied in many physical phenomena. It is used in physics to model diffusion with instantaneous jumps from one position to the next. For example, it is used in Scher and Montroll (1975) to model the transport of amorphous materials, in Montroll and Shlesinger (1984) to model the transport in disordered media, and in Weiss (1998) to model the transport in turbid media. In Helmstetter and Sornette (2002) it is used to model earthquakes. Recently, it has been used in econophysics to model the financial market, particularly to describe the movement of log prices. Examples of the application of CTRW in econophysics are discussed in Masoliver and Montero (2003), Masoliver (2006), and Scalas (2006a, 2006b).

As discussed earlier, CTRW is a random walk that permits intervals between successive walks to be independent and identically distributed. Thus, the walker starts at the point 0 at time $T_0 = 0$ and waits until time T_1 when he or she makes a jump of size θ_1, which is not necessarily positive. The walker then waits until time T_2 when he makes another jump of size θ_2, and so on. The jump sizes θ_i are also assumed to be independent and identically distributed. Thus, we assume that the times T_1, T_2, ... are the instants when the walker makes jumps. The intervals $\tau_i = T_i - T_{i-1}$, $i = 1, 2, \ldots$, are called the *waiting times* (or *pausing times*) and are assumed to be independent and identically distributed. The time at which the nth walk occurs, T_n, is given by

$$T_n = T_0 + \sum_{i=1}^{n} \tau_i \quad n = 1, 2, \ldots; \qquad T_0 = 0 \tag{3.56}$$

Thus, one major difference between the classical or discrete-time random walk (DTRW) and the CTRW is that in CTRW the waiting time between jumps is not constant as in the case of DTRW, and the step size is a random variable unlike the DTRW where the step size is ± 1.

The position of the walker at time t is given by

$$X(t) = \sum_{i=1}^{N(t)} \theta_i \qquad (3.57)$$

where the upper limit $N(t)$ is a random function of time that denotes the number of jumps up to time t and is given by

$$N(t) = \max\left\{n : T_n \leq t\right\}$$

Let T denote the waiting time and let Θ denote the jump size. Similarly, let $f_{X(t)}(x, t)$ denote the probability density function (PDF) of $X(t)$; $f_\Theta(\theta)$, the PDF of Θ; $f_T(t)$, the PDF of T; and $p_{N(t)}(n, t)$, the PMF of $N(t)$.

Let $P(x, t)$ denote the probability that the position of the walker at time t is x, given that it was in position 0 at time $t=0$; that is,

$$P(x, t) = P[X(t) = x \mid X(0) = 0] \qquad (3.58)$$

For an *uncoupled* CTRW, it is assumed that Θ and T are independent so that the joint PDF of the jump size and waiting time $f_{\Theta T}(\theta, t) = f_\Theta(\theta)f_T(t)$. In a *coupled* CTRW, Θ and T are not independent so that

$$f_\Theta(\theta) = \int_0^\infty f_{\Theta T}(\theta, t)dt \qquad (3.59)$$

$$f_T(t) = \int_{-\infty}^\infty f_{\Theta T}(\theta, t)d\theta \qquad (3.60)$$

Let $\Phi_{X(t)}(w)$ denote the characteristic function of $X(t)$, let $\Phi_\Theta(w)$ denote the characteristic function of Θ, and let $G_{N(t)}(z)$ denote the z-transform of the PMF of $N(t)$, where

$$\Phi_{X(t)}(w) = E[e^{jwX(t)}] = \int_{-\infty}^\infty e^{jwx} f_{X(t)}(x, t)dx$$

$$\Phi_\Theta(w) = E[e^{jw\Theta}] = \int_{-\infty}^\infty e^{jw\theta} f_\Theta(\theta)d\theta$$

$$G_{N(t)}(z) = E[z^{N(t)}] = \sum_{n=0}^\infty z^n p_{N(t)}(n, t)$$

Because $X(t)$ is a random sum of random variables, we know from probability theory (see Ibe (2005), for example) that

$$\Phi_{X(t)}(w) = G_{N(t)}(\Phi_\Theta(w)) \qquad (3.61)$$

Thus, if $p_{N(t)}(n, t)$ and $f_\Theta(\theta)$ are known, we can determine $\Phi_{X(t)}(w)$ and, consequently, $f_{X(t)}(x, t)$. The expected value of $X(t)$ is given by

$$E[X(t)] = E[N(t)]E[\Theta] \tag{3.62}$$

3.18.1 The Master Equation

The relationship between $P(x, t)$ and $f_{\Theta T}(\theta, t)$ is given by the following equation that is generally called the *master equation* of the CTRW:

$$P(x, t) = \delta(x)R(t) + \int_0^t \int_{-\infty}^\infty P(u, \tau)f_{\Theta T}(x - u, t - \tau)du\, d\tau \tag{3.63}$$

where $\delta(x)$ is the Dirac delta function and $R(t) = P[T>t] = 1 - F_T(t)$ is called the *survival probability*, which is the probability that the waiting time when the process is in a given state is greater than t. The equation states that the probability that $X(t)=x$ is equal to the probability that the process was in state 0 up to time t, plus the probability that the process was at some state u at time τ, where $0<\tau\le t$, and within the waiting time $t - \tau$, a jump of size $x - u$ took place. Note that

$$R(t) = \int_t^\infty f_T(v)dv = 1 - \int_0^t f_T(v)dv$$

$$f_T(t) = -\frac{dR(t)}{dt}$$

For the uncoupled CTRW, the master equation becomes

$$P(x, t) = \delta(x)R(t) + \int_0^t \int_{-\infty}^\infty P(u, \tau)f_\Theta(x - u)f_T(t - \tau)du\, d\tau \tag{3.64}$$

Let the joint Fourier-Laplace transform of $P(x, t)$ be defined as follows:

$$P^*(w, s) = \int_0^\infty e^{-st}dt \int_{-\infty}^\infty e^{jwx} P(x, t)dx \tag{3.65}$$

Then the master equation is transformed into

$$P^*(w, s) = \int_0^\infty e^{-st} \int_{-\infty}^\infty e^{jwx}\left[\delta(x)R(t) + \int_0^t \int_{-\infty}^\infty P(u, \tau)f_{\Theta T}(x - u, t - \tau)du\, d\tau\right]dx\, dt$$

$$= \tilde{R}(s) + P^*(w, s)\Phi_{\Theta T}(w, s)$$

where $\tilde{R}(s)$ is the Laplace transform of $R(t)$. This gives

$$P^*(w, s) = \frac{\tilde{R}(s)}{1 - \Phi_{\Theta T}(w, s)} \tag{3.66}$$

where $\Phi_{\Theta T}(w, s)$ is the joint Fourier-Laplace transform of $f_{\Theta T}(\theta, \tau)$. For an uncoupled CTRW, $\Phi_{\Theta T}(w, s) = \Phi_{\Theta}(w)\Phi_T(s)$. Thus,

$$P^*(w, s) = \frac{\tilde{R}(s)}{1 - \Phi_{\Theta}(w)\Phi_T(s)} \tag{3.67}$$

Also,

$$\tilde{R}(s) = \frac{1 - \Phi_T(s)}{s}$$

Thus, for the uncoupled CTRW we have that

$$P^*(w, s) = \frac{\tilde{R}(s)}{1 - \Phi_{\Theta}(w)\Phi_T(s)} = \frac{1 - \Phi_T(s)}{s\left[1 - \Phi_{\Theta}(w)\Phi_T(s)\right]} \tag{3.68}$$

Because $|\Phi_{\Theta}(w)| < 1$ if $w \neq 0$ and $|\Phi_T(s)| < 1$ if $s \neq 0$, we can rewrite Equation 3.68 as follows:

$$P^*(w, s) = \tilde{R}(s)\sum_{n=0}^{\infty}\left[\Phi_{\Theta}(w)\Phi_T(s)\right]^n \tag{3.69}$$

Taking the inverse Fourier and Laplace transforms we obtain:

$$P(x, t) = \sum_{n=0}^{\infty} P(n, t) f_{\Theta}^{(n)}(x) \tag{3.70}$$

where $P(n, t)$ is the probability that n jumps occur up to time t, and $f_{\Theta}^{(n)}(x)$ is the n-fold convolution of the number of jumps. $P(n, t)$ is given by

$$P(n, t) = \int_0^t f_T^{(n)}(t - u) R(u) \, du$$

where $f_T^{(n)}(t)$ is the n-fold convolution of the PDF of the waiting time.

Consider the special case where T is exponentially distributed with a mean of $1/\lambda$, that is,

$$f_T(t) = \lambda e^{-\lambda t}, \ t \geq 0$$

This means that

$$R_T(t) = e^{-\lambda t}, \ t \geq 0$$

In this case we have that $\Phi_T(s) = \lambda/(s+\lambda)$. Thus, for the uncoupled CTRW we obtain

$$P^*(w, s) = \frac{1 - \lambda/(s+\lambda)}{s[1 - \lambda\Phi_\Theta(w)/(s+\lambda)]} = \frac{1}{s + \lambda[1 - \Phi_\Theta(w)]} \tag{3.71}$$

Taking the inverse Laplace transform for this special case we obtain

$$P(w, t) = \exp\{-\lambda t[1 - \Phi_\Theta(w)]\} \tag{3.72}$$

Note that for the special case when $N(t)$ is a Poisson process, the CTRW becomes a compound Poisson process and the analysis is simplified as the characteristic function of $X(t)$ becomes

$$\Phi_{X(t)}(w) = \exp\{-\lambda t[1 - \Phi_\Theta(w)]\} = P(w, t) \tag{3.73}$$

which is not surprising because the fact that $N(t)$ is a Poisson process implies that T is exponentially distributed. The inverse Laplace transform of $P^*(w, s)$ for the uncoupled system is given by

$$P(w, t) = \Phi_{X(t)}(w) = G_{N(t)}(\Phi_\Theta(w)) \tag{3.74}$$

3.19   REINFORCED RANDOM WALK

Edge reinforced random walk (ERRW) was introduced by Coppersmith and Diaconis (1987) as a simple model of exploring a new city. Consider a person coming to a new city whose streets are all equally unfamiliar to him, and he randomly chooses streets along which to walk. As time goes on, the streets that have been traversed more often in the past become more familiar to him and are more likely to be traversed again.

More formally, in ERRW a random walker walks on a weighted graph $G=(V, E)$, where each edge has initial weight a. The walker moves randomly about the graph, choosing at each step one of the edges leaving from the walker's position with probabilities proportional to the weights of the edges. Whenever the walker traverses an edge, the weight of that edge is increased by one, reinforcing the likelihood of the walker traversing that edge again in the future.

For acyclic graphs, the ERRW can be modeled using Polya's urns. Thus, before we continue our discussion on ERRW we provide a brief discussion on the Polya's urn model.

3.19.1   Polya's Urn Model

Consider an urn that contains r red and b black balls. In the so-called Polya's urn model one ball is drawn randomly from the urn and its color observed; it is then placed back in the urn together with c balls of the same color, and the selection process is repeated $n-1$ times so that the total number of drawings made from the urn is n. Thus, each time a ball is selected, $c+1$ balls of the same color as the selected ball are returned to the urn. When $c=0$, we have a sampling with replacement. Thus, unlike the case of the basic model of sampling with replacement, the contents of the urn change over time, with a *self-reinforcing* property.

The Polya urn model is useful in survival analysis situations in which a system undergoes repeated stresses that are not independent of each other. For example, in the study of the spread of a contagious disease in a closed population, the more infected people there are in the population, the larger the likelihood that a noninfected person will become infected; that is, the more shocks affecting the individual. Polya's urn model has often been applied to the spread of contagious diseases.

Let R_j, $1\leq j\leq n$, denote the event that the jth ball drawn is red and let B_j, $1\leq j\leq n$, denote the event that the jth ball drawn is black. At the kth draw, there are a total of $b+r+(k-1)c$ balls in the urn. Consider the event that the first draw results in a red ball and the second draw also resulted in a red ball. Then the probability of this event is given by

$$P[R_1 \cap R_2] = P[R_1]P[R_2 \mid R_1]$$

Now,

$$P[R_1] = \frac{r}{b+r}, \quad P[R_2 \mid R_1] = \frac{r+c}{b+r+c}$$

This gives

$$P[R_1 \cap R_2] = \left\{\frac{r}{b+r}\right\}\left\{\frac{r+c}{b+r+c}\right\}$$

Similarly,

$$P[B_1 \cap R_2] = \left\{\frac{b}{b+r}\right\}\left\{\frac{r}{b+r+c}\right\}$$

Since $P[R_2] = P[R_1 \cap R_2] + P[\overline{R}_1 \cap R_2] = P[R_1 \cap R_2] + P[B_1 \cap R_2]$, the probability that the second draw results in a red ball is given by

$$P[R_2] = P[R_1 \cap R_2] + P[B_1 \cap R_2] = \left\{\frac{r}{b+r}\right\}\left\{\frac{r+c}{b+r+c}\right\} + \left\{\frac{b}{b+r}\right\}\left\{\frac{r}{b+r+c}\right\}$$

$$= \frac{r(b+r+c)}{(b+r)(b+r+c)} = \frac{r}{b+r} = P[R_1]$$

Because $P[B_2] = 1 - P[R_2]$, we have that

$$P[B_2] = \frac{b}{b+r} = P[B_1]$$

In fact, it can be shown that the probability that the kth draw results in a red ball is given by

$$P[R_k] = \frac{r}{b+r} = P[R_1] \quad k = 1, 2, \ldots$$

Another question of interest is the distribution of the number of red balls in the first n draws. Consider, for example, the sequence $I = R_1 B_2 B_3 R_4 R_5$. We have that

$$P[R_1 B_2 B_3 R_4 R_5] = \left(\frac{r}{b+r}\right)\left(\frac{b}{b+r+c}\right)\left(\frac{b+c}{b+r+2c}\right)\left(\frac{r+c}{b+r+3c}\right)\left(\frac{r+2c}{b+r+4c}\right)$$

$$= \frac{\{r(r+c)(r+2c)\}\{b(b+c)\}}{(b+r)(b+r+c)(b+r+2c)(b+r+3c)(b+r+4c)}$$

In the second line of the equation, we have rearranged the order of the factors in the numerator to make the representation appear as a product of conditional probabilities more clearly. This is the probability that five drawings result in the first three red balls and then two black ones. In fact, all possible ways of ending up with three red balls after drawing five times have equal probability. From this we can see that the probability of a particular string of R_i and B_i corresponding to a particular sequence of draws from the urn depends only on the number of R_i and B_i and not on their ordering. This property is usually called the *exchangeability* property.

3.19.2   ERRW and Polya's Urn

As stated earlier, the ERRW can be modeled using Polya's urns. At each vertex, we place an urn containing several colored balls, with each color corresponding to an edge connected to that vertex. For the vertex at which the walker starts, we initially have a balls of each color, corresponding to the initial weight of a on each edge. The first move is chosen by randomly drawing a ball from the urn, and moving along the edge corresponding to the resulting color. This is equivalent to choosing based on edge weights. Once a ball is drawn, it is put back into the urn, along with one more ball of the same color. Because the graph is acyclic, if the walker ever returns to that vertex, he will more likely do so by re-traversing that same edge. Thus, when the next decision needs to be made at that vertex, the weight of the chosen edge will have increased by exactly one, and a drawing from the urn can again be used to determine the walker's next move with the correct probabilities. All other vertices behave in the same way, except that the urns start out with a balls of each color, plus one extra ball of the color corresponding to the edge that goes toward the walker's starting position: the first time the walker arrives at that vertex, it will do so by that edge, and the weight will already be $a+1$.

3.19.3   ERRW Revisited

Consider a graph $G=(V,E)$ in which the edges are given positive weights. At time 0 the weights are nonrandom; edge e has weight $a_e>0$, $e \in E$. We denote by $w_n(e)$ the weight of edge e at time n (just after the nth step) and by $w_n(v)$ the sum of the weights of the edges incident to vertex v.

We define reinforced random walk with starting point $v_0 \in V$ to be a sequence X_0, Y_1, X_1, Y_2, X_2, ... with $X_i \in V$, $Y_i \in E$ and $e_i = \{X_{i-1}, X_i\}$ for all $i \in N = \{1, 2, \ldots\}$. Furthermore $P[X_0 = v_0] = 1$ and

$$P[Y_{n+1} = e, X_{n+1} = v \mid X_0, Y_1, X_1, \ldots, Y_n, X_n] = \begin{cases} \dfrac{w_n(e)}{w_n(X_n)} & \text{if } e = \{X_n, v\} \\ 0 & \text{otherwise} \end{cases}$$

and the weights satisfy

$$w_0(e) = a_e$$

$$w_{n+1}(e) = \begin{cases} w_n(e)+1 & \text{if } Y_{n+1} = e \\ w_n(e) & \text{otherwise} \end{cases}$$

The *node-reinforced random walk* can be similarly defined. One algorithm that can be used is as follows. Recall that $\Gamma(i)$ is the set of the neighbors of node i. A record is kept for the number of visits from each node to each of its neighbors. For node k, let $w_k^j(i)$ denote the number of visits to node $i \in \Gamma(k)$ at the jth time node k is visited. The probability that node $i \in \Gamma(k)$ is chosen as the next hop is

$$p_i = \frac{w_k^j(i)}{\sum_{i \in \Gamma(k)} w_k^j(i)} \quad i \in \Gamma(k)$$

After the choice of the next hop has been made, the weight of the selected node is updated according to the rule: $w_k^{j+1}(i) = w_k^j(i)+1$. The weights of the other nodes are unchanged. As in the ERRW, the walker will tend to revisit nodes already visited. We assume that the initial weight assignment is $w_k^0(i) = 1$ for all i and k.

One fundamental question about the reinforced random walk concerns recurrence: Do almost all paths of the edge reinforced random walk visit all vertices infinitely often? This is still an open question. More detailed discussion on reinforced random walks can be found in Davis (1990) and Pemantle (2007).

3.20   MISCELLANEOUS RANDOM WALK MODELS

There are several random walk models that space does not permit us to discuss in great detail. In this section we summarize some of these models. They include

1. Geometric Random Walk
2. Gaussian Random Walk
3. Random Walk with Memory

3.20.1   Geometric Random Walk

The geometric random walk is used in asset modeling. Let P_t denote the price of an asset at time t. The net return over holding the asset from time $t-1$ to time t is the profit during the holding period divided by the value of the asset at the beginning of the holding period, which is given by

$$R_t = \frac{P_t - P_{t-1}}{P_{t-1}} = \frac{P_t}{P_{t-1}} - 1 \Rightarrow \frac{P_t}{P_{t-1}} = 1 + R_t$$

where it is understood that a negative value of R_t signifies a loss. If we define the gross return over the most recent k periods by $P_t/P_{t-k} = 1 + R_t(k)$, then we may express this ratio as follows:

$$\frac{P_t}{P_{t-k}} = 1 + R_t(k) = \left(\frac{P_t}{P_{t-1}} \right) \left(\frac{P_{t-1}}{P_{t-2}} \right) \cdots \left(\frac{P_{t-k+1}}{P_{t-k}} \right)$$
$$= (1 + R_t)(1 + R_{t-1}) \cdots (1 + R_{t-k+1})$$

If we define the log price at time t by $p_t = \log(P_t)$, then the log return at time t, r_t, is given by

$$r_t = \log \left(\frac{P_t}{P_{t-1}} \right) = \log(1 + R_t) = p_t - p_{t-1}$$

then we have that the k-period log return is given by

$$r_t(k) = \log(1 + R_t(k)) = \log\{(1 + R_t)(1 + R_{t-1}) \cdots (1 + R_{t-k+1})\}$$
$$= \log(1 + R_t) + \log(1 + R_{t-1}) + \cdots + \log(1 + R_{t-k+1})$$
$$= r_t + r_{t-1} + \cdots + r_{t-k+1}$$

Since a random walk is a mathematical description of a trajectory that consists of taking random steps, $r_t(k)$ is essentially a random walk model. Also, we have that

$$\frac{P_t}{P_{t-k}} = 1 + R_t(k) = \exp\{r_t(k)\} = \exp\{r_t + r_{t-1} + \cdots + r_{t-k+1}\}$$

If we define $k = t$, we obtain

$$P_t = P_0 \exp\{r_t + r_{t-1} + \cdots + r_1\}$$

The process $\{P_t, t \geq 0\}$ whose logarithm is a random walk is called a geometric random walk, which is also called an *exponential random walk*.

3.20.2 Gaussian Random Walk

A Gaussian random walk is the discrete-time process

$$S_n = \sum_{k=1}^{n} X_k$$

where S_n is the location of a walker relative to the origin after n steps such that the kth step size X_k is a normally distributed random variable. If the random variables X_1, X_2, ..., X_n are independent and identically distributed, then S_n is called a simple Gaussian random walk. If the PDF of each step X_k depends on a function of the previous step X_{k-1}, denoted by $\mu(X_{k-1})$, the random walk is called a correlated Gaussian random walk. One example of a correlated Gaussian random walk is one in which the PDF of X_k is given by

$$f_{X_k}(x_k) = \frac{1}{\sqrt{2\pi\sigma^2}} \exp\left(\frac{-|x_k - \mu(x_{k-1})|^2}{2\sigma^2}\right) \quad k = 1, 2, \ldots, n$$

3.20.3 Random Walk with Memory

A random walk has memory m if it cannot at any transition visit a site visited in the last m steps. A random walk with memory 0 is just a standard random walk. A random walk with memory 1 is a random walk without backtracking (or nonreversing random walk). A random walk with infinite memory is a self-avoiding random walk. Both the nonreversing random walk and the self-avoiding random walk are discussed in Chapter 4.

A first-in first-out list, called the memory list M, is maintained. The list contains the node identifiers n_k of the last M nodes visited during the random walk; that is, $M = \{n_1, n_2, \ldots, n_M\}$. At each node, the next hop is chosen uniformly among the neighbors of the node that are not currently in the memory list M. The use of memory eliminates loops in random walks. One-hop loops can be prevented by using $M = 1$, both two-hop and one-hop loops can be prevented by using $M = 2$, and so on. Thus, if

N is the number of nodes in the network a memory of $M=N-1$ totally eliminates loops in the random walk. However, the introduction of memory $(M \geq 1)$ in random walks can lead to a deadlock in which the walker cannot move because all neighboring nodes are on the memory list.

3.21 SUMMARY

This chapter has been devoted to discussing the basic concepts on random walk, particularly one-dimensional motion. These concepts include the gambler's ruin problem, applications of the reflection principle, random walk on a graph, random walk and electric networks, CRW, CTRW, and reinforced random walk. We also provided a brief discussion on exponential random walk, Gaussian random walk, and random walk with memory. More information is provided in Chapter 4 on two-dimensional random walk.

PROBLEMS

3.1 A bag contains four red balls, three blue balls, and three green balls. Jim plays a game in which he bets $1 to draw a ball from the bag. If he draws a red ball, he wins $1; otherwise he loses $1. Assume that the balls are drawn with replacement and that Jim starts with $50 with the hope of reaching $100 at which point he stops playing. However, if he loses all his money before this, the game also ends and he becomes bankrupt. What is the probability that the game ends with Jim being $50 richer than he had at the beginning?

3.2 Mark and Kevin play a series of games of cards. During each game each player bets $1, and whoever wins the game gets the $2. Sometimes a game ends in a tie in which case neither player loses his money. Mark is a better player than Kevin and has a probability 0.5 of wining each game, a probability 0.3 of losing each game, and a probability 0.2 of tying with Kevin. Initially Mark had $9 and Kevin had $6, and the game is over when either player is bankrupt.

    1. Give the state-transition diagram of the process.

    2. If r_k denotes the probability that Mark is ruined, given that the game is currently in state k, obtain an expression for r_k in the first game when the process is in state k.

3.3 Chris has $20 and Dana has $30. They decide to play a game in which each pledges $1 and flips a fair coin. If both coins come up on the same side, Chris wins the $2, and if they come up on different sides, Dana wins the $2. The game ends when either of them has all the money. What is the probability that Chris wins the game?

3.4 Consider the random walk $S_n = X_1 + X_2 + \cdots + X_n$, where the X_i are independent and identically distributed Bernoulli random variables that take on the value 1 with probability $p=0.6$ and the value -1 with probability $q=1-p=0.4$.

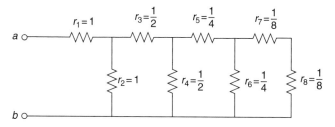

FIGURE 3.17   Figure for Problem 3.7.

1. Find the probability $P[S_8=0]$.
2. What value of p maximizes $P[S_8=0]$?

3.5   Let N denote the number of times that an asymmetric random walk that takes a step to the right with probability p and a step to the left with probability $q=1-p$ revisits its starting point. Show that the PMF of N is given by

$$p_N(n) = P[N = n] = \beta(1-\beta)^n \quad n = 0, 1, \ldots$$

where $\beta=|p-q|$.

3.6   Consider an asymmetric random walk that takes a step to the right with probability p and a step to the left with probability $q=1-p$. Assume that there are two absorbing barriers, a and b, and that the walk starts at the point k, where $b<k<a$.

1. What is the probability that the walk stops at b?
2. What is the mean number of steps until the walk stops?

3.7   Find the effective resistance between nodes a and b and the escape probability of the electric network shown in Figure 3.17.

3.8   Consider a CRW with stay. That is, a walker can move to the right, to the left, or not move at all. Given that the move in the current step is to the right, then in the next step it will move to the right again with probability a, to the left with probability b and remain in the current position with probability $1-a-b$. Given that the walker did not move in the current step, then in the next step it will move to the right with probability c, to the left with probability d, and not move again with probability $1-c-d$. Finally, given that the move in the current step is to the left, then in the next step it will move to the right with probability g, to the left again with probability h, and remain in the current position with probability $1-g-h$. Let the process be represented by the bivariate process $\{X_n, Y_n\}$, where X_n is the location of the walker after n steps and Y_n is the nature of the nth step (i.e., right, left, or no move). Let π_1 be the limiting probability that the process is in "right" state, π_0 the limiting probability that it is in the "no move" state, and π_{-1} the limiting probability that it is in the "left" state, where $\pi_1+\pi_0+\pi_{-1}=1$. Let $\Pi=\{\pi_1, \pi_0, \pi_{-1}\}$.

    1. Find the values of π_1, π_0, and π_{-1}.

    2. Obtain the transition probability matrix of the process.

3.9 Consider a CTRW $\{X(t)|t \geq 0\}$ in which the jump size, Θ, is normally distributed with mean μ and variance σ^2, and the waiting time, T, is exponentially distributed with mean $1/\lambda$. Obtain the master equation, $P(x, t)$, which is the probability that the position of the walker at time t is $X(t)=x$, given that it was in position 0 at time $t=0$.

CHAPTER 4

TWO-DIMENSIONAL RANDOM WALK

4.1 INTRODUCTION

In Chapter 3 we discussed the one-dimensional (1D) random walk. In this chapter, we extend the discussion to the two-dimensional (2D) random walk, which has been used in many ecological studies to study the movement of animals in Codling (2003), Codling et al. (2008), and Othmer et al. (1988), as well as in polymer science. We consider five types of 2D random walk, which are as follows:

1. Pearson random walk
2. Symmetric 2D random walk
3. Alternating 2D random walk
4. Self-avoiding random walk
5. Nonreversing 2D random walk

In the Pearson random walk, the walker takes a step length of length L_1 in a direction that is uniformly distributed between 0 and 2π. Then he takes a second step of length L_2 at another random angle, and so on. Thus, if the walker is at location Y_{n-1} at the end of the $(n-1)th$ step, the next location will be a point on a circle of radius L_{n-1} centered at Y_{n-1}. This is the only non-lattice walk among the four models listed above.

Elements of Random Walk and Diffusion Processes, First Edition. Oliver C. Ibe.
© 2013 John Wiley & Sons, Inc. Published 2013 by John Wiley & Sons, Inc.

In the symmetric random walk (SRW), the walker takes one step either in the positive x direction, negative x direction, positive y direction, or negative y direction with equal probability. The step size is fixed, and each step is independent of the other steps.

In alternating random walk (ARW), the walker takes the first step in positive x or negative x direction and next step in the positive y or negative y direction. That is, a walker's alternate steps are in x and y directions resulting in a diagonal walk. The step size is fixed, each step is independent of the other steps, and two consecutive Bernoulli trials govern the outcome of a complete step. The outcome of the first trial is used to decide whether to go in the positive x or negative x direction, and the outcome of the second trial is used to decide whether to go in the positive y or negative y direction.

As stated in Chapter 3, both the self-avoiding random walk (SAW) and the nonreversing are examples of random walk with memory. In a self-avoiding walk, the walker is not allowed to visit a previously visited site. Once he reaches a site where the next step gets him to a previously visited location, the walk stops, and we say that the walker is trapped. Thus, the SAW has an infinite memory, which enables it to avoid already visited sites.

In the nonreversing random walk (NRRW), which has memory 1, the walker chooses his next direction with equal probability excluding the direction that will get him immediately back to the location from where he came to the current location. That is, immediate backtracking is not allowed. Thus, unlike the self-avoiding walk, he can go visit a previously visited location, but not immediately after leaving the location. Both the SAW and NRRW are examples of a restricted random walk.

Figure 4.1 illustrates three examples of different random walk models on the lattice. Figure 4.1a illustrates the SRW where the walker can go east, west, north, or south after each step with equal probability. Initially, the walker is at location $(0, 0)$. At the next step, he goes east to location $(1, 0)$. Next he goes north to location $(1, 1)$, and so on. In this case, the walk is memoryless. Figure 4.1b illustrates an ARW where a step in east or west is followed by a step in north or south, but never east and west. As stated earlier, this is a way to model a diagonal walk. Finally, Figure 4.1c illustrates an NRRW. Here, we assume that the walker is initially in location $(0, 0)$. Assume he next goes east to location $(1, 0)$, then at the next step, he cannot go west

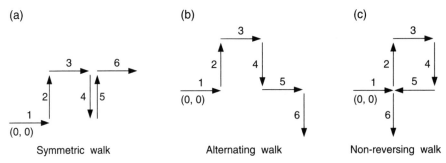

FIGURE 4.1   Examples of different types of 2D lattice random walk.

since this will bring him back to the location from where he immediately came to the current location. Thus, he can either go east, north, or south. Assume that he goes north to location $(1, 1)$. Then, the next step cannot be south since this will bring him back to location $(1, 0)$. So, he can go west, east, or north, and so on. As stated earlier, he can return to a previously visited location, but never immediately after leaving the location.

4.2   THE PEARSON RANDOM WALK

As discussed earlier, a Pearson random walk consists of a sequence of straight lines each at a random angle with respect to the preceding one. We consider a 2D random walk in which the walker takes a step of length L_1 in a direction that makes an angle θ_1 with the x-axis. Then at the end of the step he chooses a random direction that makes an angle θ_2 with the x-axis and takes a step of length L_2, and so on. We assume that the step lengths are drawn from a random variable L that has a probability density function (PDF) $f_L(l)$, $l \geq 0$, with a finite mean and a finite variance. We also assume that the turn angles are uniformly distributed between 0 and 2π; that is, the PDF of the random variable Θ that represents the turn angles is

$$f_\Theta(\theta) = \frac{1}{2\pi} \quad 0 \leq \theta \leq 2\pi \tag{4.1}$$

4.2.1   Mean-Square Displacement

We can represent a single step length by the projections along the x and y axes. Thus, we define x_i and y_i as the x-component and y-component of the ith step, respectively. That is,

$$x_i = L_i \cos \theta_i$$
$$y_i = L_i \sin \theta_i$$

Let X_n denote the x-component of the displacement of the walker after n steps and let Y_n denote the y-component of the displacement of the walker after n steps. If R_n is the total displacement after n steps, we have that

$$R_n^2 = X_n^2 + Y_n^2$$
$$L_i = \sqrt{x_i^2 + y_i^2}$$

Since Θ is uniformly distributed, $E[X_n] = E[Y_n] = 0$. This means that there are as many negative x-components as positive of equal magnitude so that they cancel out, and there are as many negative y-components as positive of equal magnitude so that they also cancel out. Also, $E[R_n^2] = E[X_n^2] + E[Y_n^2]$, where

$$E\left[X_n^2\right] = E\left[\left(\sum_{i=1}^{n}x_i\right)^2\right] = E\left[\left(\sum_{i=1}^{n}x_i\right)\left(\sum_{j=1}^{n}x_j\right)\right] = E\left[\sum_{i=1}^{n}x_i^2 + 2\sum_{i>j}x_ix_j\right]$$

$$= \sum_{i=1}^{n}E\left[x_i^2\right] + 2\sum_{i>j}E[x_ix_j]$$

$$E\left[Y_n^2\right] = E\left[\left(\sum_{i=1}^{n}y_i\right)^2\right] = E\left[\left(\sum_{i=1}^{n}y_i\right)\left(\sum_{j=1}^{n}y_j\right)\right] = E\left[\sum_{i=1}^{n}y_i^2 + 2\sum_{i>j}y_iy_j\right]$$

$$= \sum_{i=1}^{n}E\left[y_i^2\right] + 2\sum_{i>j}E[y_iy_j]$$

$$E\left[R_n^2\right] = \sum_{i=1}^{n}\left\{E\left[x_i^2 + y_i^2\right]\right\} + 2\sum_{i>j}E\left[x_ix_j\right] + 2\sum_{i>j}E\left[y_iy_j\right]$$

$$= nE\left[L^2\right] + 2\sum_{i>j}E\left[x_ix_j\right] + 2\sum_{i>j}E\left[y_iy_j\right]$$

Now, $E[x_ix_j] = E[L_iL_j\cos(\Theta_i)\cos(\Theta_j)] = (E[L])^2 \ E[\cos(\Theta_i)\cos(\Theta_j)] = 0$ for $i \neq j$ and $E[y_iy_j] = (E[L])^2 \ E[\sin(\Theta_i)\sin(\Theta_j)] = 0$ for $i \neq j$. Thus,

$$E\left[R_n^2\right] = nE\left[L^2\right] \tag{4.2}$$

The distribution of Θ can be chosen to permit a correlated random walk by biasing the walk in some directions more than others. Thus, rather than its being uniformly distributed as in Equation 4.1, we assume some biased distribution. For this case we define

$$E[L] = \int_0^\infty lf_L(l)dl$$

$$E[L^2] = \int_0^\infty l^2 f_L(l)dl$$

$$c = \int_0^{2\pi} \cos(\theta)f_\Theta(\theta)d\theta \tag{4.3}$$

$$s = \int_0^{2\pi} \sin(\theta)f_\Theta(\theta)d\theta \tag{4.4}$$

In Kareiva and Shigesada (1983), it is shown that for the total displacement after n consecutive moves, R_n, we have that

$$E\left[R_n^2\right] = nE[L^2] + 2\left(E[L]\right)^2 \left\{ \frac{n(c-c^2-s^2)-c}{(1-c)^2+s^2} + \gamma \frac{2s^2+(c+s^2)^{(n+1)/2}}{\left[(1-c)^2+s^2\right]^2} \right\} \quad (4.5)$$

where

$$\gamma = [(1-c)^2 - s^2]\cos\left\{(n+1)\alpha\right\} - 2s(1-c)\sin\left\{(n+1)\alpha\right\}$$
$$\alpha = \arctan(s/c)$$

When Θ has a uniform distribution, $c=s=0$ and we have that $E\left[R_n^2\right]=nE\left[L^2\right]$ as we obtained earlier.

4.2.2 Probability Distribution

Let $Q_n(r)$ denote the probability of the walker being at location r after n steps, and let $P_n(x)$ denote the probability of the walker being at location x after n steps. If we define the PDF of the component of the step length along the x-axis by $f_X(x)$, then applying the Chapman–Kolmogorov equation we have that

$$P_n(x) = \int_{-\infty}^{\infty} P_{n-1}(u) f_X(x-u)du \quad (4.6)$$

This equation states that the probability of having the walker at x after n steps is equal to the probability of his arriving at u in $n-1$ steps and making up the difference $x-u$ in one additional step. Let $P_n^*(w)$ denote the characteristic function of $P_n(x)$ and let $F_X^*(w)$ be the characteristic function of $f_X(x)$. Then, from the convolution nature of Equation 4.6 we obtain

$$P_n^*(w) = P_{n-1}^*(w) F_X^*(w)$$

If we assume that the walker started at the origin so that $P_0(x) = \delta(x)$, then we obtain

$$P_n^*(w) = P_{n-2}^*(w)\left\{F_X^*(w)\right\}^2 = \cdots = \left\{F_X^*(w)\right\}^n \quad (4.7)$$

From this we can obtain $P_n(x)$ by the inverse transform

$$P_n(x) = \frac{1}{2\pi}\int_{-\infty}^{\infty} P_n^*(w)e^{-iwx}dw = \frac{1}{2\pi}\int_{-\infty}^{\infty}\left\{F_X^*(w)\right\}^n e^{-iwx}dw \quad (4.8)$$

Note that we only know the PDF $f_L(l)$ of step lengths and the PDF $f_\Theta(\theta)$ of the turn angles. Thus, we need to obtain the PDF $f_X(x)$ whose characteristic function appears in the equation. As stated earlier, the x-component of the ith step is given by $x_i = l_i \cos \theta_i$. Thus, the random variable X can be defined by

$$X = L \cos \Theta = LV$$

where $V = \cos \Theta$. As shown in Ibe (2005), if U and W are two continuous random variables, the PDF of their product $Z = UW$ is given by

$$f_Z(z) = \int_{-\infty}^{\infty} \frac{1}{|y|} f_{UW}\left(y, \frac{z}{y}\right) dy$$

Since L and Θ are independent, we have that

$$f_X(x) = \int_{-\infty}^{\infty} \frac{1}{|w|} f_L(w) f_V\left(\frac{x}{w}\right) dw$$

We know that the PDF of V is given by

$$f_V(v) = \frac{1}{\pi\sqrt{1-v^2}}, \quad -1 < v < 1$$

Thus, we obtain

$$f_X(x) = \frac{1}{\pi}\int_{-\infty}^{\infty} \frac{1}{|w|} f_L(w)\left\{\frac{|w|}{\sqrt{w^2-x^2}}\right\} dw = \frac{1}{\pi}\int_{x}^{\infty} f_L(w)\left\{\frac{1}{\sqrt{w^2-x^2}}\right\} dw \quad (4.9)$$

The characteristic function of $f_X(x)$ becomes

$$F_X^*(w) = \int_{-\infty}^{\infty} f_X(x) e^{iwx} dx = \frac{1}{\pi}\int_{-\infty}^{\infty}\int_{x}^{\infty} f_L(u)\left\{\frac{1}{\sqrt{u^2-x^2}}\right\} e^{iwx} du dx$$

In the special case when L is fixed, say $L=k$, we obtain $f_L(l) = \delta(l-k)$, where $\delta(x)$ is the Dirac delta function and thus,

$$f_X(x) = \frac{1}{\pi}\int_{x}^{\infty} f_L(w)\left\{\frac{1}{\sqrt{w^2-x^2}}\right\} dw = \frac{1}{\pi}\int_{x}^{\infty} \delta(w-k)\left\{\frac{1}{\sqrt{w^2-x^2}}\right\} dw$$

$$= \frac{1}{\pi}\left\{\frac{1}{\sqrt{k^2-x^2}}\right\} \quad -k < x < k$$

$$F_X^*(w) = \int_{-\infty}^{\infty} f_X(x) e^{iwx} dx = \frac{1}{\pi}\int_{-\infty}^{\infty}\left\{\frac{1}{\sqrt{k^2-x^2}}\right\} e^{iwx} dx dw$$

Another method of solution is to recognize that we can use the transformation

$$f_X(x)dx = f_\Theta(\theta)d\theta = \frac{d\theta}{2\pi}$$

Thus, for the special case of constant step length, we have that

$$F_X^*(w) = \int_{-\infty}^{\infty} f_X(x)e^{iwx}dx = \frac{1}{2\pi}\int_0^{2\pi} e^{iwk\cos\theta}d\theta = J_0(wk) \tag{4.10}$$

where $J_0(wk)$ is the Bessel function of the first kind of order zero. Thus, the characteristic function of $P_n(x)$ is given by

$$P_n^*(w) = \{J_0(wk)\}^n \tag{4.11}$$

From this we obtain $P_n(x)$ as follows:

$$P_n(x) = \frac{1}{2\pi}\int_{-\infty}^{\infty} P_n^*(w)e^{-iwx}dw = \frac{1}{2\pi}\int_{-\infty}^{\infty} e^{-iwx}\{J_0(wk)\}^n dw$$

$$= \frac{1}{\pi}\int_0^{\infty} \cos(wx)\{J_0(wk)\}^n dw \tag{4.12}$$

The last equality is due to the fact that $J_0(wk)$ is even in w. An algorithm that can numerically calculate integrals of power laws multiplied by a product of Bessel functions of the first kind quickly with very small error has been developed in Van Duen and Cools (2006, 2008).

For the radial distribution we have that

$$Q_n(r) = \int_{-\infty}^{\infty} Q_{n-1}(u)f_R(r-u)du$$

From this we obtain

$$Q_n^*(w) = Q_{n-1}^*(w)F_R^*(w) = Q_{n-2}^*(w)\{F_R^*(w)\}^2 = \cdots = \{F_R^*(w)\}^n$$

where we assume that the walker started at the origin so that $Q_0(r) = \delta(r)$ where r is a 2D vector. Also, for a 2D isotropic walk, we have that

$$f_R(r) = \frac{1}{2\pi|r|}\delta(|r|-k)$$

To compute the characteristic function of $f_R(r)$ we introduce polar coordinates (ρ, ϕ), where $|r| = \rho$, the parameter w is a 2D vector and the dot product $r \cdot w = \rho|w|\cos\phi$. Thus,

$$F_R^*(w) = \frac{1}{2\pi\rho}\int_0^{2\pi}\left\{\int_0^\infty \rho e^{i\rho|w|\cos\phi}\delta(\rho - k)d\rho\right\}d\phi = \frac{1}{2\pi}\int_0^{2\pi}\left\{\int_0^\infty e^{i\rho|w|\cos\phi}\delta(\rho - k)d\rho\right\}d\phi$$

$$= \frac{1}{2\pi}\int_0^{2\pi} e^{i|w|k\cos\phi}d\phi = J_0(|w|k)$$

We can obtain $Q_n(r)$ by the inverse transform

$$Q_n(r) = \frac{1}{(2\pi)^2}\int Q_n^*(w)\exp(-iw \cdot r)d^2w = \frac{1}{(2\pi)^2}\int\left\{F_R^*(w)\right\}^n \exp(-iw \cdot r)d^2w$$

$$= \frac{1}{(2\pi)^2}\int\left\{J_0(|w|k)\right\}^n \exp(-iw \cdot r)d^2w$$

We also introduce polar coordinates (u, φ) in the w-plane and obtain

$$Q_n(r) = \frac{1}{2\pi}\int_0^\infty u\left\{J_0(|u|k)\right\}^n\left\{\frac{1}{2\pi}\int_0^{2\pi}\exp(-iu|r|\cos\varphi)d\varphi\right\}du$$

$$= \frac{1}{2\pi}\int_0^\infty u\left\{J_0(|u|k)\right\}^n J_0(ur)du \tag{4.13}$$

We can use the same method discussed earlier to obtain the approximate solution.

4.3   THE SYMMETRIC 2D RANDOM WALK

A 2D random walk is a stochastic process $\{(X, Y), n = 0, 1, \ldots\}$ with state space $\Omega = \{(i, j)|i, j \in \{0, \pm1, \pm2, \ldots\}\}$. Let $P_k(x, y)$ denote the probability that after k steps, the walker, starting from the origin, is at location (x, y). Assume that at the end of this step he has taken m_1 steps in the positive x-direction, m_2 steps in the negative x-direction, n_1 steps in the positive y-direction, and n_2 steps in the negative y-direction. Let p_1 denote the probability that the walker takes a step in the positive x-direction, p_2 be the probability that he takes a step in the negative x-direction, q_1 be the probability that he takes a step in the positive y-direction, and q_2 be the probability that he takes a step in the negative y-direction, where $p_1 + p_2 + q_1 + q_2 = 1$. $P_k(x, y)$ is given by the multinomial distribution:

$$P_k(x, y) = \begin{pmatrix} k \\ m_1 \quad m_2 \quad n_1 \quad n_2 \end{pmatrix} p_1^{m_1} p_2^{m_2} q_1^{n_1} q_2^{n_2} \tag{4.14}$$

where

$$m_1 - m_2 = x \Rightarrow m_1 = x + m_2$$
$$n_1 - n_2 = y \Rightarrow n_1 = y + n_2$$

The total number of steps is

$$m_1 + m_2 + n_1 + n_2 = k$$

Assume that the walker has taken a total of l steps in the x direction. Thus, he takes $(k-l)$ steps in the y direction. We have that

$$m_1 + m_2 = l$$
$$n_1 + n_2 = k - l$$

From this we obtain

$$x = m_1 - (l - m_1) \Rightarrow m_1 = \frac{l+x}{2}, \quad m_2 = \frac{l-x}{2}$$
$$y = n_1 - (k - l - n_1) \Rightarrow n_1 = \frac{k-l+y}{2}, \quad n_2 = \frac{k-l-y}{2}$$

Therefore, we obtain

$$P_k(x, y) = \sum_{l=|x|}^{k-y} \left(\frac{k}{\dfrac{l+x}{2} \quad \dfrac{l-x}{2} \quad \dfrac{k-l+y}{2} \quad \dfrac{k-l-y}{2}} \right) p_1^{\frac{l+x}{2}} p_2^{\frac{l-x}{2}} q_1^{\frac{k-l+y}{2}} q_2^{\frac{k-l-y}{2}}$$

For a SRW, the probability of going in any direction is the same. Thus, we have that

$$P_k(x, y) = \left(\frac{1}{4}\right)^k \sum_{l=|x|}^{k-y} \left(\frac{k}{\dfrac{l+x}{2} \quad \dfrac{l-x}{2} \quad \dfrac{k-l+y}{2} \quad \dfrac{k-l-y}{2}} \right)$$

$$= \left(\frac{1}{4}\right)^k \sum_{l=|x|}^{k-y} \left[\frac{k!}{\left(\dfrac{l+x}{2}\right)!\left(\dfrac{l-x}{2}\right)!\left(\dfrac{k-l+y}{2}\right)!\left(\dfrac{k-l-y}{2}\right)!} \right]$$

$$= \left(\frac{1}{4}\right)^k \sum_{l=|x|}^{k-y} \left\{ \frac{k! l! (k-l)!}{l!(k-l)!\left(\dfrac{l+x}{2}\right)!\left(\dfrac{l-x}{2}\right)!\left(\dfrac{k-l+y}{2}\right)!\left(\dfrac{k-l-y}{2}\right)!} \right\}$$

$$= \left(\frac{1}{4}\right)^k \sum_{l=|x|}^{k-y} \left\{ \binom{k}{l} \binom{l}{\dfrac{l+x}{2}} \binom{k-l}{\dfrac{k-l+y}{2}} \right\} \tag{4.15}$$

Unfortunately, we cannot easily evaluate $P_k(x, y)$ even for small to moderate values of k. However, we can obtain the return probability $P_k(0, 0)$ for a symmetric walk as follows: In order to return to the origin, the walker would take an even number of steps, say $2n$, such that he takes l steps to the right, l steps to the left, $n-l$ steps up, and $n-l$ steps down. Thus, we have that

$$P_{2n}(0,0) = \left(\frac{1}{4}\right)^{2n} \sum_{l=0}^{n}\left\{\frac{(2n)!}{(l)!(l)!(n-l)!(n-l)!}\right\} = \left(\frac{1}{4}\right)^{2n}\sum_{l=0}^{n}\left\{\frac{(2n)!n!n!}{(l)!(l)!(n-l)!(n-l)!n!n!}\right\}$$

$$= \left(\frac{1}{4}\right)^{2n}\sum_{l=0}^{n}\binom{2n}{n}\binom{n}{l}\binom{n}{l} = \left(\frac{1}{4}\right)^{2n}\binom{2n}{n}\sum_{l=0}^{n}\binom{n}{l}^2$$

But

$$\sum_{l=0}^{n}\binom{n}{l}^2 = \binom{2n}{n}$$

This means that

$$P_{2n}(0,0) = \left(\frac{1}{4}\right)^{2n}\binom{2n}{n}^2 = \frac{1}{4^{2n}}\left(\frac{(2n)!}{n!n!}\right)^2 \tag{4.16}$$

4.3.1 Stirling's Approximation of Symmetric Walk

From the Stirling's formula we have that for large values of n,

$$n! \sim \sqrt{2\pi n}\left(\frac{n}{e}\right)^n = \sqrt{2\pi n}\, n^n e^{-n}$$

Thus, for large values of n we have that

$$P_{2n}(0,0) = \frac{1}{4^{2n}}\left(\frac{(2n)!}{n!n!}\right)^2 \sim \frac{1}{4^{2n}}\left[\frac{\sqrt{2\pi}(2n)^{2n+1/2}e^{-2n}}{\left(\sqrt{2\pi}(n)^{n+1/2}e^{-n}\right)^2}\right]^2 = \frac{1}{4^{2n}}\left[\frac{(2n)^{2n}\sqrt{4\pi n}}{n^{2n}(2\pi n)}\right]^2$$

$$= \frac{1}{4^{2n}}\left[\frac{4^n}{\sqrt{\pi n}}\right]^2$$

$$= \frac{1}{\pi n} \tag{4.17}$$

4.3.2   Probability of Eventual Return for Symmetric Walk

We are interested in the probability of eventual return to the origin of a random walker in a symmetric 2D random walk. As discussed earlier,

$$P_{2n}(0,0) = P[S_{2n} = (0,0)] = \frac{1}{4^{2n}}\left(\frac{(2n)!}{n!n!}\right)^2 \approx \frac{1}{\pi n}$$

Now, let the indicator variable I_n be defined as follows:

$$I_n = \begin{cases} 1 & \text{if } S_{2n} = (0,0) \\ 0 & \text{otherwise} \end{cases}$$

Thus, $E[I_n] = P[S_{2n} = (0,0)]$. Let N denote the total number of returns to the origin, which is sometimes called the *number of equalizations*; that is,

$$N = \sum_{n=1}^{\infty} I_n$$

Therefore,

$$E[N] = E\left[\sum_{n=1}^{\infty} I_n\right] = \sum_{n=1}^{\infty} E[I_n]$$

$$= \sum_{n=1}^{\infty} P[S_{2n} = (0,0)] = \sum_{n=1}^{\infty} \frac{1}{4^{2n}}\left(\frac{(2n)!}{n!n!}\right)^2$$

$$\approx \frac{1}{\pi}\sum_{n=1}^{\infty} \frac{1}{n}$$

$$= \infty$$

where the last equality follows from the fact that the Riemann's zeta function $\sum_{n=1}^{\infty} 1/n^k$ diverges for $k \le 1$. This result implies that the walk is a recurrent process and the probability of eventual return to the origin infinitely many times is 1. Another way to interpret the result is based on the Borel–Cantelli lemma, which can be stated as follows:

Lemma: Let $\{E_n\}$ be a sequence of independent events in some probability space. If the sum of the probabilities of E_n is infinite (i.e., $\sum_{n=1}^{\infty} P[E_n] = \infty$), then the probability that the event occurs infinitely many times is 1. If the sum of the probabilities of E_n is finite, then the probability that the event occurs infinitely many times is 0.

Thus, according to this lemma, since $\sum_{n=1}^{\infty} P\left[S_{2n} = (0,0)\right] = \infty$, the walker returns infinitely many times to the origin with probability 1.

4.3.3 Mean-Square Displacement

In Chapter 3, we saw that the mean displacement for the 1D random walk with step length of δ is given by

$$E\left[S_N^2\right] = N\delta^2 + N(N-1)(2p\delta - \delta)^2$$

where p is the probability of taking a step in the positive direction. For a SRW where $p = 0.5$, we have that $E\left[S_N^2\right] = N\delta^2$. By defining a completion time $t = N\tau$ for the N steps and a diffusion coefficient $D = \delta^2/2\tau$, where τ is the time it takes to complete one step, we obtained $E\left[S_N^2(t)\right] = 2Dt$. Consider a symmetric 2D random walk in which the walker arrives at the point (x, y). The square displacement of the point (x, y) from the origin is $S_{XY}^2 = S_X^2 + S_Y^2$, where S_X is the displacement in the horizontal direction and S_Y is the displacement in the vertical direction. Thus, $E\left[S_{XY}^2\right] = E\left[S_X^2\right] + E\left[S_Y^2\right]$. If $E\left[S_X^2(t)\right] = 2Dt$, then $E\left[S_Y^2(t)\right] = 2Dt$. Therefore, we have that

$$E\left[S_{XY}^2(t)\right] = 4Dt \tag{4.18}$$

4.3.4 Two Independent Symmetric 2D Random Walkers

We extend the problem of two random walkers we discussed in Chapter 3 for the 1D random walk to 2D SRW. Specifically we consider two walkers who start at the origin at time 0 and independently execute 2D SRWs. We are interested in the probability that they meet again after each has taken N steps.

As in the 1D random walk, for them to meet again they would have to take the same number of steps to the north, the same number of steps to the south, the same number of steps to the east, and the same number of steps to the west. Again, the property of exchangeability enables us to consider only the number steps in each direction and not the order of these steps as long as the number is the same for both walkers. Assume that each of them takes k north–south steps and $N-k$ east–west steps. Additionally, out of the k north–south steps, assume that each takes $y \leq k$ steps to the north; and out of the $N-k$ east–west steps, assume that each of them takes $r \leq N-k$ steps to the right. Thus, the probability that they meet again after N steps is given by

$$P_N = \sum_{k=0}^{N}\sum_{r=0}^{N-k}\sum_{y=0}^{k}\left\{\binom{k}{y}\binom{N-k}{r}\frac{1}{4^N}\right\}^2 = \frac{1}{4^{2N}}\sum_{k=0}^{N}\sum_{r=0}^{N-k}\sum_{y=0}^{k}\left\{\binom{k}{y}\binom{N-k}{r}\right\}^2$$

$$= \frac{1}{4^{2N}}\sum_{k=0}^{N}\sum_{r=0}^{N-k}\binom{N-k}{r}^2\sum_{y=0}^{k}\binom{k}{y}^2 = \frac{1}{4^{2N}}\sum_{k=0}^{N}\binom{2k}{k}\binom{2N-2k}{N-k}$$

$$= \frac{1}{4^{2N}}\binom{2N}{N} = \frac{(2N)!}{(4^N N!)^2}$$

This result is similar in structure to that obtained for the 1D random walk.

4.4 THE ALTERNATING RANDOM WALK

In the ARW, the walker takes the first step in positive x or negative x direction and next step in the positive y or negative y direction. That is, a walker's alternate steps are in x and y directions. The step size is fixed, each step is independent of the other steps, and two consecutive Bernoulli trials govern the outcome of a complete step. The outcome of the first trial is used to decide positive x or negative x direction, and the outcome of the second trial is used to decide positive y or negative y direction. Thus, ARW is fundamentally a diagonal walk.

Let $P_l(x, y)$ denote the probability that after l steps, the walker, starting from the origin, is at location (x, y). Since to get to a location the walker needs to make two complete moves, we require that l be an even number. Assume that at the end of this step the walker has taken m_1 steps in the positive x-direction, m_2 steps in the negative x-direction, n_1 steps in the positive y-direction, and n_2 steps in the negative y-direction. Let p_1 denote the probability that the walker takes a step in the positive x-direction, p_2 be the probability that he takes a step in the negative x-direction, q_1 be the probability that he takes a step in the positive y-direction, and q_2 be the probability that he takes a step in the negative y-direction, where $p_1+p_2+q_1+q_2=1$. $P_{2k}(x, y)$ is given by the multinomial distribution:

$$P_{2k}(x, y) = \binom{2k}{m_1 \ m_2 \ n_1 \ n_2} p_1^{m_1} p_2^{m_2} q_1^{n_1} q_2^{n_2} \tag{4.19}$$

For the ARW, at each trial the walker has two choices to make. For example, in the first trial, the survivor takes steps in $+x$ or $-x$ direction and in the following trial he takes $+y$ or $-y$ direction. Each trial is independent of x and y and the number of steps taken in horizontal direction must be equal to the number of steps in vertical direction. Thus, this is equivalent to a diagonal random walk. If the walker takes $2k$ steps from the origin to get to location (x, y), then he has taken k steps in horizontal direction and k steps in vertical direction. Thus, we have that

$$m_1 + m_2 = k$$
$$n_1 + n_2 = k$$

Solving for m_1, m_2, n_1, and n_2 gives

$$x = m_1 - (k - m_1) \Rightarrow m_1 = \frac{k+x}{2}, \quad m_2 = \frac{k-x}{2}$$

$$y = n_1 - (k - n_1) \Rightarrow n_1 = \frac{k+y}{2}, \quad n_2 = \frac{k-y}{2}$$

Thus, for the 2D ARW, the multinomial distribution becomes

$$P_{2k}(x, y) = \begin{pmatrix} 2k \\ \dfrac{k+x}{2} \quad \dfrac{k-x}{2} \quad \dfrac{k+y}{2} \quad \dfrac{k-y}{2} \end{pmatrix} p_1^{\frac{k+x}{2}} p_2^{\frac{k-x}{2}} q_1^{\frac{k+y}{2}} q_2^{\frac{k-y}{2}}$$

In the case where the probability of reaching any valid location is equal, the equation simplifies to

$$P_{2k}(x, y) = \left(\frac{1}{4}\right)^{2k} \left\{ \frac{(2k)!}{\left(\dfrac{k+x}{2}\right)! \left(\dfrac{k-x}{2}\right)! \left(\dfrac{k+y}{2}\right)! \left(\dfrac{k-y}{2}\right)!} \right\}$$

$$= \left(\frac{1}{4}\right)^{2k} \left\{ \frac{(2k)!\,k!\,k!}{k!\,k!\left(\dfrac{k+x}{2}\right)! \left(\dfrac{k-x}{2}\right)! \left(\dfrac{k+y}{2}\right)! \left(\dfrac{k-y}{2}\right)!} \right\} \qquad (4.20)$$

$$= \left(\frac{1}{4}\right)^{2k} \binom{2k}{k} \binom{k}{\dfrac{k+x}{2}} \binom{k}{\dfrac{k+y}{2}}$$

where $k+x$ and $k+y$ are required to be even numbers. Note that a "valid location" is one where the walker has taken the same number of steps in both orthogonal axes.

The return probability is obtained by observing that it takes multiples of four steps to return to the origin. Thus,

$$P_{4n}(0, 0) = \left(\frac{1}{4}\right)^{4n} \binom{4n}{2n} \binom{2n}{n}^2 = \left(\frac{1}{4}\right)^{4n} \left\{ \frac{(4n)!}{(n!)^4} \right\} \qquad (4.21)$$

4.4.1 Stirling's Approximation of Alternating Walk

Applying the Stirling's approximation for large values of n to Equation 4.21 we obtain

$$P_{4n}(0,0) \sim \left(\frac{1}{4}\right)^{4n} \frac{\sqrt{2\pi(4n)}\left(\dfrac{4n}{e}\right)^{4n}}{\left\{\sqrt{2\pi n}\left(\dfrac{n}{e}\right)^{n}\right\}^{4}} = \frac{\sqrt{8\pi n}}{\left\{\sqrt{2\pi n}\right\}^{4}} \qquad (4.22)$$

$$= \frac{1}{\pi^{3/2} n^{3/2} \sqrt{2}}$$

4.4.2 Probability of Eventual Return for Alternating Walk

From (4.22) we observe that

$$\sum_{n=1}^{\infty} P_{4n}(0,0) \approx \frac{1}{\pi^{3/2}\sqrt{2}} \sum_{n=1}^{\infty} \frac{1}{n^{1.5}} < \infty$$

where the inequality follows from the fact that the Riemann's zeta function $\sum_{n=1}^{\infty} 1/n^{k}$ converges if $k>1$. Thus, according to the Borel–Cantelli lemma, the probability that the walker returns to the origin infinitely many times is 0, which means that the walk is a transient process.

4.5 SELF-AVOIDING RANDOM WALK

As defined earlier, a SAW is a form of correlated random walk in which the walker is not allowed to visit a previously visited site. This type of walk is typically rendered on a lattice and serves as a model for linear polymer molecules. Polymers are the fundamental building blocks in biological systems. A polymer is a long chain of monomers (i.e., groups of atoms) joined to one another by chemical bonds. These polymer molecules form together randomly, with the restriction that no overlaps can occur. Thus, as also discussed earlier, SAW has an infinite memory.

One of the problems with SAW is that although it is a model of random walk, its trajectory cannot be described in terms of transition probabilities. Thus, it is a non-Markovian process and consequently it is mathematically difficult to analyze. Currently, its analysis is carried out either through simulation or a direct enumeration of a small number of steps.

It is possible for the random walker to get into a *trap*; that is, he reaches a site whose neighbors have already been visited, at which point the walk ends. The SAW in Figure 4.2 illustrates such a self-avoiding walk called *self-trapping walk* where any movement from the last site the walker visits will result in revisiting a previous site. At this point the walk will terminate.

FIGURE 4.2   Example of self-trapping walk.

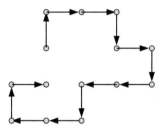

FIGURE 4.3   Example of self-avoiding polygon.

A related type of SAW is the *self-avoiding polygon* (SAP), which is a closed self-avoiding walk on a lattice. This means that SAP is a SAW whose final site is adjacent to the first site, as illustrated in Figure 4.3.

There are several reasons for the interest in SAW. First, it is an interesting mathematical concept that is regarded as a classical combinatorial problem. Also, its origin lies in polymer science where it represents a realistic model of long-chain polymers in dilute solution. Polymers are long chain molecules consisting of a large number of monomers held together by chemical bonds. The sclf-avoiding property reflects the fact that two monomers cannot occupy the same point in space. One measure of a polymer is number of monomers (i.e., its length); another measure is the average distance from one end to the other. Thus, while SAW is an interesting mathematical concept, it is an important tool in polymer science.

Two issues that are usually addressed in the analysis of SAW are the following:

1. How many possible self-avoiding walks of length n can be found?
2. When does the walker get trapped?

With respect to the first question, let C_n denote the number of self-avoiding walks of n steps that begin at the origin. Many attempts have been made by several authors to obtain C_n. Table 4.1 shows some values of C_n for different values of n on 2D square lattice as given in Slade (1994). Values of C_n for n as high as 39 have been obtained.

Because of the intractability of the problem the preferred method of solution is by simulation. Bounds on C_n can be obtained as follows. The number of SRWs in a 2D lattice of length n is 4^n since at each site the walker can choose from four possible directions. But for a self-avoiding walk, he can choose from four directions in the first step, and thereafter he chooses from no more than three other locations. Thus, an

TABLE 4.1   Values of C_n on Two-Dimensional Square Lattice, $n \leq 24$

n	C_n	n	C_n
1	4	13	881,500
2	12	14	2,374,444
3	36	15	6,416,596
4	100	16	17,245,332
5	284	17	46,466,676
6	780	18	124,658,732
7	2,172	19	335,116,620
8	5,916	20	897,697,164
9	16,258	21	2,408,806,028
10	44,100	22	6,444,560,484
11	120,292	23	17,266,613,812
12	324,932	24	46,146,397,316

upper bound on C_n is $C_n \leq 4(3^{n-1})$. A lower bound is obtained by noting that one way to avoid a previously visited location is by moving only in the positive direction that involves only two choices at each location. Thus, we have that

$$2^n \leq C_n \leq 4\left(3^{n-1}\right) \tag{4.23}$$

These bounds seem to suggest that we can express C_n by

$$C_n = \alpha(n)\mu^n \tag{4.24}$$

for some positive number $\mu > 0$ and

$$\lim_{n \to \infty}\left[\alpha(n)\right]^{1/n} = 1.$$

This is the so-called Hammersely–Morton theorem, see Hughes (1995). Any self-avoiding walk of $m+n$ steps can be decomposed into a self-avoiding walk of m steps followed by a self-avoiding walk of n steps. Thus, C_n satisfies the submultiplicative inequality

$$C_{m+n} \leq C_m C_n \tag{4.25}$$

Note that the reverse inequality is not true as concatenating two self-avoiding walks does not always yield a self-avoiding walk. If we define $\beta_n = \log C_n$, then β_n satisfies the "sub-additive inequality"

$$\beta_{m+n} \leq \beta_m + \beta_n \tag{4.26}$$

Just as we do not know exactly how many n-step self-avoiding walks there are that start at the origin, we do not know exactly the answer to the second question, which deals with when a self-avoiding walker gets trapped. In the absence of exact analytical methods, asymptotic estimation from exact enumeration of shorter walks has been used to study this property of SAWs. The most important statistic is the mean-square length of the walk. Let R_n denote the endpoint of a SAW. The mean-square length is $E\left[R_n^2\right]$, which has been shown by many studies to be of the form

$$E\left[R_n^2\right] = An^\nu \qquad n \to \infty$$

where $\nu = 1.5$ for a 2D SAW and A is a constant; see Barber and Ninham (1970) and Hughes (1995).

With respect to when a walker becomes trapped, let $P_T(n)$ denote the probability that a self-avoiding walker is trapped after exactly n steps. In Hemmer and Hemmer (1984), it is shown that

$$P_T(n) \propto (n-6)^{0.6} e^{-n/40} \quad n > 6 \tag{4.27a}$$

$$P_T(n) = 0 \qquad n \leq 6 \tag{4.27b}$$

and the average length of self-avoiding walks before the walker is trapped is about 71. Thus, when $n \leq 6$, the probability of being trapped is negligible. Traps begin to appear from the seventh step onward. In the 2D self-avoiding walk, there are eight different walks that form traps at the seventh step. Let K_n denote the number of SAWs that become trapped on the nth step. Table 4.2 shows the values of K_n for different values of n.

TABLE 4.2   Values of K_n on Two-Dimensional Square Lattice, $7 \leq n \leq 14$

n	K_n
7	8
8	16
9	88
10	200
11	760
12	1,824
13	6,016
14	14,880

Figure 4.4 shows all the eight different walks that form traps at the seventh step.

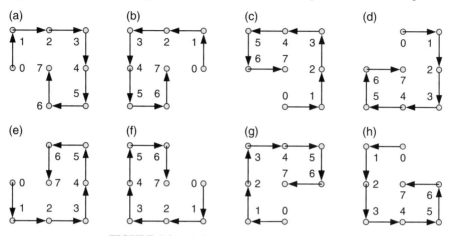

FIGURE 4.4 Walks trapped at the seventh step.

Figure 4.5 shows 4 of the 16 different walks that form traps at the eighth step.

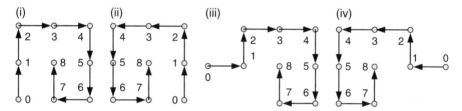

FIGURE 4.5 Some of the walks trapped at the eighth step.

4.6 NONREVERSING RANDOM WALK

The NRRW is an intermediate stage between a classical random walk and a self-avoiding walk. It is a correlated random walk that allows the walker to go left, right, or forward at each intersection, but not to make a U-turn and go back to the location from where he just came. Thus, on a square lattice a nonreversing walker has four choices for the first step, but only three choices for each step thereafter. The model was first introduced by Temperley (1956). Another way to view NRRW is as follows. A classical random walk is a walk with no memory while SAW is a walk with unlimited memory because the walker has to remember all the sites he has visited. NRRW lies between a walk with no memory and a walk with unlimited memory; there are bounds on the amount of memory possessed by a walker because he only avoids the last site visited. Thus, as we discussed earlier, NRRW is a random walk with memory 1.

One reason for the study of NRRW is the following. The symmetric 2D random walk assumes that each step is taken independently of the previous steps. This is probably the case for nonhuman walkers but not for humans. A human walker is not likely to go back immediately to the location from where he came, which suggests that his walk has some memory. Therefore, a more reasonable model is one in which the walker will not immediately go back to the location from where he came to the current location. This means that, if the previous step was a west-to-east walk, the next step cannot be an east-to-west walk and vice versa. Similarly, if the previous step was a north-to-south walk, the next step cannot be a south-to-north walk, and vice versa. We assume that if the walker came to the current location by performing a west-to-east movement, he will choose to go east with probability 1/3, north with probability 1/3, or south with probability 1/3, and so for other directions.

Note that in a self-avoidance walk, the walker can never visit a previously visited location. However, in the NRRW the walker can visit a previously visited location, except that he cannot do so immediately after leaving the location since immediate U-turns are forbidden. He can revisit a previously visited location only through visits to at least one other location before doing so. In other words, a visit to a previously visited location is possible via a sequence of locations that loops back to the location.

The 1D NRRW is a trivial walk in which movement is along one direction only. That is, there is no backtracking and once the initial direction has been chosen the walker keeps moving in the same direction. If we assume that the walker has left the origin (or equivalently, the process is in the steady state), the 2D case can be modeled in two ways. First, it can be viewed as a random walk in a 2D lattice in which the turn angle Θ takes on three discrete values; that is, the PMF of Θ is

$$p_{\Theta}(\theta) = \begin{cases} \frac{1}{3} & \theta = -\frac{\pi}{2} \\ \frac{1}{3} & \theta = 0 \\ \frac{1}{3} & \theta = \frac{\pi}{2} \end{cases} \tag{4.28}$$

where angular measurements are made relative to the direction of the last motion. We can also define the distribution of the turn angle by the PDF

$$f_{\Theta}(\theta) = \frac{1}{3}\left\{ \delta\left(\theta + \frac{\pi}{2}\right) + \delta\left(\theta - \frac{\pi}{2}\right) + \delta(\theta) \right\} \tag{4.29}$$

In this way we can analyze the model in the same way that we did for the Pearson random walk.

The second method is to model it by a Markov chain, as discussed in Narayanan (2011). Let U be the state that represents the event that the last walk was an upward step, D the state that represents the event that the last walk was a downward step, R the state that represents the event that the last walk was a step to the right, and L the state that represents the event that the last walk was a step to the left. Then Figure 4.6 is the Markov chain for the walk. For example, when it is in state U, it cannot make a transition to state D because this will cause it to return the immediate past state, but it can with equal probability go to U, L, or R and so on for the other states.

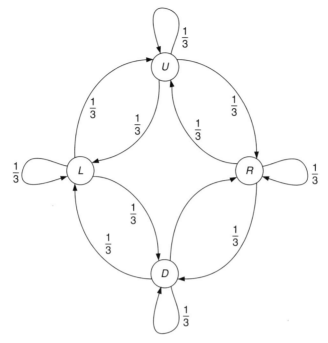

FIGURE 4.6   Markov chain for the NRRW.

If we denote state R by 1, state U by 2, state L by 3 and state D by 4, then the transition probability matrix for the process is given by

$$P = \begin{bmatrix} \frac{1}{3} & \frac{1}{3} & 0 & \frac{1}{3} \\ \frac{1}{3} & \frac{1}{3} & \frac{1}{3} & 0 \\ 0 & \frac{1}{3} & \frac{1}{3} & \frac{1}{3} \\ \frac{1}{3} & 0 & \frac{1}{3} & \frac{1}{3} \end{bmatrix} \qquad (4.30)$$

which is a doubly stochastic matrix. With this matrix we can compute the different parameters of interest. For example, we can obtain P^n since it is the inverse z-transform of the matrix $[I - Pz]^{-1}$, (see Ibe (2005)), where I is the identity matrix.

As is well known, P^n consists of two sets of components. One component is a constant term C that has the property that all its rows are identical and the elements of each row are the limiting-state probabilities. The second component $T(n)$ is a transient term. Thus, if $T(z)$ is the z-transform of $T(n)$, we have that

$$[I - Pz]^{-1} = \frac{1}{1-z}C + T(z) \qquad (4.31)$$

Therefore, to obtain P^n we proceed as follows:

$$I - Pz = \begin{bmatrix} 1-\frac{z}{3} & -\frac{z}{3} & 0 & -\frac{z}{3} \\ -\frac{z}{3} & 1-\frac{z}{3} & -\frac{z}{3} & 0 \\ 0 & -\frac{z}{3} & 1-\frac{z}{3} & -\frac{z}{3} \\ -\frac{z}{3} & 0 & -\frac{z}{3} & 1-\frac{z}{3} \end{bmatrix}$$

$$[I - Pz]^{-1} = \frac{A}{1 - \dfrac{4z}{3} + \dfrac{2z^2}{9} + \dfrac{4z^3}{27} - \dfrac{z^4}{27}} = \frac{A}{(1-z)\left(1+\dfrac{z}{3}\right)\left(1-\dfrac{z}{3}\right)^2}$$

where

$$A = \begin{bmatrix} 1-z+\dfrac{z^2}{9}+\dfrac{z^3}{27} & \dfrac{z}{3}-\dfrac{2z^2}{9}+\dfrac{z^3}{27} & \dfrac{2z^2}{9}-\dfrac{2z^3}{27} & \dfrac{z}{3}-\dfrac{2z^2}{9}+\dfrac{z^3}{27} \\ \dfrac{z}{3}-\dfrac{2z^2}{9}+\dfrac{z^3}{27} & 1-z+\dfrac{z^2}{9}+\dfrac{z^3}{27} & \dfrac{z}{3}-\dfrac{2z^2}{9}+\dfrac{z^3}{27} & \dfrac{2z^2}{9}-\dfrac{2z^3}{27} \\ \dfrac{2z^2}{9}-\dfrac{2z^3}{27} & \dfrac{z}{3}-\dfrac{2z^2}{9}+\dfrac{z^3}{27} & 1-z+\dfrac{z^2}{9}+\dfrac{z^3}{27} & \dfrac{z}{3}-\dfrac{2z^2}{9}+\dfrac{z^3}{27} \\ \dfrac{z}{3}-\dfrac{2z^2}{9}+\dfrac{z^3}{27} & \dfrac{2z^2}{9}-\dfrac{2z^3}{27} & \dfrac{z}{3}-\dfrac{2z^2}{9}+\dfrac{z^3}{27} & 1-z+\dfrac{z^2}{9}+\dfrac{z^3}{27} \end{bmatrix}$$

Thus,

$$[I - Pz]^{-1} = \frac{1}{1-z}\begin{bmatrix} \frac{1}{4} & \frac{1}{4} & \frac{1}{4} & \frac{1}{4} \\ \frac{1}{4} & \frac{1}{4} & \frac{1}{4} & \frac{1}{4} \\ \frac{1}{4} & \frac{1}{4} & \frac{1}{4} & \frac{1}{4} \\ \frac{1}{4} & \frac{1}{4} & \frac{1}{4} & \frac{1}{4} \end{bmatrix} + \frac{1}{1+\frac{z}{3}}\begin{bmatrix} \frac{1}{4} & -\frac{1}{4} & \frac{1}{4} & -\frac{1}{4} \\ -\frac{1}{4} & \frac{1}{4} & -\frac{1}{4} & \frac{1}{4} \\ \frac{1}{4} & -\frac{1}{4} & \frac{1}{4} & -\frac{1}{4} \\ -\frac{1}{4} & \frac{1}{4} & -\frac{1}{4} & \frac{1}{4} \end{bmatrix}$$

$$+ \frac{1}{1-\frac{z}{3}}\begin{bmatrix} \frac{1}{2} & 0 & -\frac{1}{2} & 0 \\ 0 & \frac{1}{2} & 0 & -\frac{1}{2} \\ -\frac{1}{2} & 0 & \frac{1}{2} & 0 \\ 0 & -\frac{1}{2} & 0 & \frac{1}{2} \end{bmatrix} + \frac{1}{\left(1-\frac{z}{3}\right)^2}\begin{bmatrix} 0 & 0 & 0 & 0 \\ 0 & 0 & 0 & 0 \\ 0 & 0 & 0 & 0 \\ 0 & 0 & 0 & 0 \end{bmatrix}$$

From this we obtain

$$
P^n = \begin{bmatrix} \frac{1}{4} & \frac{1}{4} & \frac{1}{4} & \frac{1}{4} \\ \frac{1}{4} & \frac{1}{4} & \frac{1}{4} & \frac{1}{4} \\ \frac{1}{4} & \frac{1}{4} & \frac{1}{4} & \frac{1}{4} \\ \frac{1}{4} & \frac{1}{4} & \frac{1}{4} & \frac{1}{4} \end{bmatrix} + (-\tfrac{1}{3})^n \begin{bmatrix} \frac{1}{4} & -\frac{1}{4} & \frac{1}{4} & -\frac{1}{4} \\ -\frac{1}{4} & \frac{1}{4} & -\frac{1}{4} & \frac{1}{4} \\ \frac{1}{4} & -\frac{1}{4} & \frac{1}{4} & -\frac{1}{4} \\ -\frac{1}{4} & \frac{1}{4} & -\frac{1}{4} & \frac{1}{4} \end{bmatrix} + (\tfrac{1}{3})^n \begin{bmatrix} \frac{1}{2} & 0 & -\frac{1}{2} & 0 \\ 0 & \frac{1}{2} & 0 & -\frac{1}{2} \\ -\frac{1}{2} & 0 & \frac{1}{2} & 0 \\ 0 & -\frac{1}{2} & 0 & \frac{1}{2} \end{bmatrix}
$$

$$
+ (n+1)(\tfrac{1}{3})^n \begin{bmatrix} 0 & 0 & 0 & 0 \\ 0 & 0 & 0 & 0 \\ 0 & 0 & 0 & 0 \\ 0 & 0 & 0 & 0 \end{bmatrix} \quad n = 0, 1, 2, \ldots
$$

$$
= \begin{bmatrix} \frac{1}{4} & \frac{1}{4} & \frac{1}{4} & \frac{1}{4} \\ \frac{1}{4} & \frac{1}{4} & \frac{1}{4} & \frac{1}{4} \\ \frac{1}{4} & \frac{1}{4} & \frac{1}{4} & \frac{1}{4} \\ \frac{1}{4} & \frac{1}{4} & \frac{1}{4} & \frac{1}{4} \end{bmatrix} + (-\tfrac{1}{3})^n \begin{bmatrix} \frac{1}{4} & -\frac{1}{4} & \frac{1}{4} & -\frac{1}{4} \\ -\frac{1}{4} & \frac{1}{4} & -\frac{1}{4} & \frac{1}{4} \\ \frac{1}{4} & -\frac{1}{4} & \frac{1}{4} & -\frac{1}{4} \\ -\frac{1}{4} & \frac{1}{4} & -\frac{1}{4} & \frac{1}{4} \end{bmatrix} + (\tfrac{1}{3})^n \begin{bmatrix} \frac{1}{2} & 0 & -\frac{1}{2} & 0 \\ 0 & \frac{1}{2} & 0 & -\frac{1}{2} \\ -\frac{1}{2} & 0 & \frac{1}{2} & 0 \\ 0 & -\frac{1}{2} & 0 & \frac{1}{2} \end{bmatrix}
$$

$$\tag{4.32}$$

Also,

$$
\lim_{n \to \infty} P^n = \begin{bmatrix} \frac{1}{4} & \frac{1}{4} & \frac{1}{4} & \frac{1}{4} \\ \frac{1}{4} & \frac{1}{4} & \frac{1}{4} & \frac{1}{4} \\ \frac{1}{4} & \frac{1}{4} & \frac{1}{4} & \frac{1}{4} \\ \frac{1}{4} & \frac{1}{4} & \frac{1}{4} & \frac{1}{4} \end{bmatrix} = C
$$

which is the matrix associated with the root $1 - z$; this means that the limiting state probabilities are given by

$$
\pi = \begin{bmatrix} \frac{1}{4} & \frac{1}{4} & \frac{1}{4} & \frac{1}{4} \end{bmatrix}
$$

The reason why the entries of π are equal is because P is a doubly stochastic matrix. Note that when $n=0$ we obtain the identity matrix, and when $n=1$ we obtain P. The last two matrices in Equation 4.32 constitute the transient term $T(n)$ of P^n, and it can be observed that each row of $T(n)$ sums to zero.

Knowledge of P^n does not help us to directly compute the return probability. However, it enables us to determine such a parameter as the occupancy time, $\phi_{ik}(n)$, which is the expected time it takes the process to visit state k in the first n transitions, given that it starts in state i. If we denote the matrix of the $\phi_{ik}(n)$ by $\Phi(n)$, then it can be shown that

$$\Phi(n) = \left[\phi_{ik}(n)\right] = \sum_{r=0}^{n} P^r \tag{4.33}$$

Let R_N denote the total displacement from the origin after N steps. The mean-square displacement of this walk after the walker has taken N steps is given in Verdier and DiMarzio (1969) as

$$E\left[R_N^2\right] = 2N - 1.5\left[1 - \left(\frac{1}{3}\right)^N\right] \tag{4.34}$$

4.7 EXTENSIONS OF THE NRRW

We briefly discuss two extension of the NRRW. These are:

1. Noncontinuing random walk (NCRW)
2. Nonreversing and noncontinuing random walk (NRNCRW)

4.7.1 The Noncontinuing Random Walk

This is the complement of the NRRW in the sense that in the NCRW continuation in the direction of the preceding step is not allowed. Thus, at the end of each step, the walker can go back to the last location from where he came to the current location or take a step in either direction of the perpendicular axis. The Markov chain for this work is shown in Figure 4.7. In this case, self-loops are eliminated, and the transition probability matrix is still a doubly stochastic matrix, as in the NRRW.

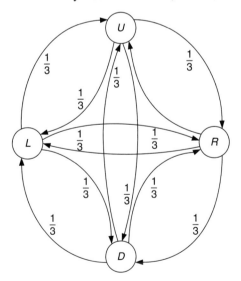

FIGURE 4.7 Markov chain for the NCRW.

The state transition probability matrix is given by

$$
P = \begin{bmatrix} 0 & \frac{1}{3} & \frac{1}{3} & \frac{1}{3} \\ \frac{1}{3} & 0 & \frac{1}{3} & \frac{1}{3} \\ \frac{1}{3} & \frac{1}{3} & 0 & \frac{1}{3} \\ \frac{1}{3} & \frac{1}{3} & \frac{1}{3} & 0 \end{bmatrix} \tag{4.35}
$$

With this we can analyze the model in a manner similar to that for the NRRW. The mean-square end-to-end length of this walk after the walker has taken N steps is given in Verdier and DiMarzio (1969) as

$$
E\left[R_N^2 \right] = 0.5N + 0.375\left[1 - \left(-\tfrac{1}{3} \right)^N \right] \tag{4.36}
$$

4.7.2   The Nonreversing and Noncontinuing Random Walk

This model combines the nonreversing and noncontinuing properties. Thus, after each step the random walker can only go to one of the two directions of the perpendicular axis. This means that there are only two choices available to the walker at the end of each step. This model is essentially the ARW that we discussed earlier in this chapter. The Markov chain for this model is shown in Figure 4.8.

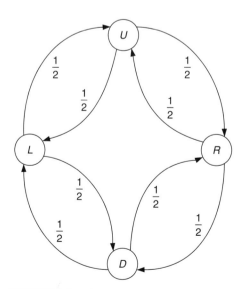

FIGURE 4.8   Markov chain for the NRNCRW.

The state transition probability matrix, which is still a doubly stochastic matrix, is given by

$$P = \begin{bmatrix} 0 & \frac{1}{2} & 0 & \frac{1}{2} \\ \frac{1}{2} & 0 & \frac{1}{2} & 0 \\ 0 & \frac{1}{2} & 0 & \frac{1}{2} \\ \frac{1}{2} & 0 & \frac{1}{2} & 0 \end{bmatrix} \tag{4.37}$$

Note that the definition of state in this model is different from the definition of state in the ARW that we analyzed earlier. Here we are concerned with the direction of motion while in the ARW model we defined a state as the coordinates of the exact location of the walker.

4.8 SUMMARY

This chapter has been devoted to 2D random walks. Five different random walk models have been discussed, which are the Pearson random walk, SRW, ARW, SAW, and the NRRW. We also discussed two extensions of the NRRW. In one case, the walker is forbidden to go in the same direction that brought him to the current location, and the other case he is forbidden to go back or continue in the same direction. Different properties of these models have been discussed. Any interested reader may refer to the references for more detailed information on any one of these models.

CHAPTER 5

BROWNIAN MOTION

5.1 INTRODUCTION

Brownian motion is a stochastic process that has applications in fields as vast and different as economics, biology, and management science. Mathematically, it can be thought of as a continuous-time process in which over every infinitely small time interval Δt, the entity under consideration moves one "step" in a certain direction. This suggests that Brownian motion can be viewed as a "random walk" process, as we will demonstrate shortly.

The physical manifestation of Brownian motion was observed by the Scottish botanist Robert Brown in 1827. His interpretation of this process was based on the movement of small pollen particles suspended in a drop of water. In his experiments, the pollen particles appeared to move in a completely random fashion, stumping Brown and his colleagues. Upon further investigation, Brown and others verified that this phenomenon was not unique to pollen particles, but rather was exhibited by many different types of microscopic particles suspended in a fluid.

Brown's initial observations of Brownian motion went almost 80 years without any significant mathematical or even physical explanation until 1905 when Einstein published a paper that claimed that the motion of these microscopic particles stemmed from the constant forces exerted on the particles from the surrounding fluid, thanks to individual fluid molecules bumping into the particles, thus sending them in motion. Since the pollen grains were completely surrounded by these fluid molecules, they

Elements of Random Walk and Diffusion Processes, First Edition. Oliver C. Ibe.
© 2013 John Wiley & Sons, Inc. Published 2013 by John Wiley & Sons, Inc.

experience these forces in every conceivable direction, which explains why the particles do not move in a set pattern or direction.

The actual development of Brownian motion as a stochastic process did not surface until 1923 when Norbert Weiner, an MIT mathematician, established the modern mathematical framework of what is known today as the Brownian motion random process. This is why Brownian motion is sometimes referred to as the *Wiener process*. In fact, the study of Brownian motion today mostly involves the stochastic process pioneered by Weiner, rather than the physical process studied by Brown.

We can gain some intuition to the behavior of Brownian motion by comparing it with a simple random walk. Specifically, assume that a random walk $X(t)$ increases by an infinitesimal step size Δx over each infinitesimal time increment Δt with probability p and decreases by Δx with probability $1-p$. This allows the process $X(t)$ to be treated as a continuous function of t in which the change in x is without discontinuities. Then for each step we have that

$$E[X(t+\Delta t)-X(t)]=p\Delta x+(1-p)(-\Delta x)=(2p-1)\Delta x$$
$$E\left[\left\{X(t+\Delta t)-X(t)\right\}^2\right]=p(\Delta x)^2+(1-p)(-\Delta x)^2=(\Delta x)^2$$

The number of time units of duration Δt in the time t is $[t/\Delta t]$, where $[t/\Delta t]$ is the largest integer that is less than or equal to $t/\Delta t$. Thus, the location of $X(t)$ at time t is given by the sum of $[t/\Delta t]$ random variables each of which has a value Δx with probability p and a value $-\Delta x$ with probability $1-p$. The expected value of the sum of these random variables is the mean of their sum, which is $[t/\Delta t](2p-1)\Delta x$. Since these random variables are independent and identically distributed, the variance of their sum is $[t/\Delta t](\Delta x)^2$.

If this process is to approximate the Brownian motion $W(t)\sim N(\mu t,\sigma^2 t)$, we must have that in the limit

$$\left[\frac{t}{\Delta t}\right](2p-1)\Delta x=\mu t$$
$$\left[\frac{t}{\Delta t}\right](\Delta x)^2=\frac{(\Delta x)^2}{\Delta t}t=\sigma^2 t$$

If for large t we approximate $[t/\Delta t]$ by $t/\Delta t$, then from these two equations we obtain

$$\Delta x=\sigma\sqrt{\Delta t}$$
$$p=\frac{1}{2}\left(1+\frac{\mu\Delta t}{\Delta x}\right)=\frac{1}{2}\left(1+\frac{\mu}{\sigma}\sqrt{\Delta t}\right)$$

Let $T>0$ be some fixed interval and let $\Delta t=T/n$, where $n=1, 2, \dots$. Assume that the random variables Y_k, $k=1, \dots, n$, independently take values $\sigma\sqrt{\Delta t}$ with probability

$p = \left\{1 + \left(\mu\sqrt{\Delta t}\right)/\sigma\right\}/2$ and $-\sigma\sqrt{\Delta t}$ with probability $1 - \left\{1 + \left(\mu\sqrt{\Delta t}\right)/\sigma\right\}/2$.

Define S_n by

$$S_n = \sum_{k=1}^{n} Y_k$$

Then from the central limit theorem, $S_n \to W(t)$ as $n \to \infty$. Thus, the Brownian motion can be regarded as a random walk defined over an infinitesimally small step size Δx and infinitesimally small time intervals Δt between walks such that both Δx and Δt go to 0 in such a way that $(\Delta x)^2/\Delta t$ remains constant.

More formally, the Brownian motion $\{W(t), t \geq 0\}$ is a stochastic process that models random continuous motion. It is considered to be the continuous-time analog of the random walk and can also be considered as a continuous-time Gaussian process with independent increments. In particular, the Brownian motion has the following properties:

1. $W(0) = 0$; that is, it starts at 0.
2. $W(t)$ is continuous in $t \geq 0$; that is, it has continuous sample paths with no jumps.
3. It has both stationary and independent increments.
4. For $0 \leq s < t$, the random variable $W = W(t) - W(s)$ has a normal distribution with mean 0 and variance $\sigma_W^2 = \sigma^2(t - s)$. That is, $W \sim N(0, \sigma^2(t-s))$.

Brownian motion is an important building block for modeling continuous-time stochastic processes. In particular, it has become an important framework for modeling financial markets. The path of a Brownian motion is always continuous, but it is nowhere smooth; consequently, it is nowhere differentiable. The fact that the path is continuous means that a particle in Brownian motion cannot jump instantaneously from one point to another.

Because $W(0) = 0$, then according to property 4,

$$W(t) = W(t) - W(0) \sim N(0, \sigma^2(t - 0)) = N(0, \sigma^2 t) \tag{5.1}$$

Thus, $W(t-s) \sim N(0, \sigma^2(t-s))$; that is, $W(t) - W(s)$ has the same distribution as $W(t-s)$. This also means that another way to define a Brownian motion $\{W(t), t \geq 0\}$ is that it is a process that satisfies conditions 1, 2, and 3 along with the condition $W(t) \sim N(0, \sigma^2 t)$.

There are many reasons for studying the Brownian motion. As stated earlier, it is an important building block for modeling continuous-time stochastic processes because many classes of stochastic processes contain Brownian motion. It is a Markov process, a Gaussian process, a martingale, a diffusion process, as well as a Levy process. Over the years it has become a rich mathematical object. For example, it is the central theme of stochastic calculus.

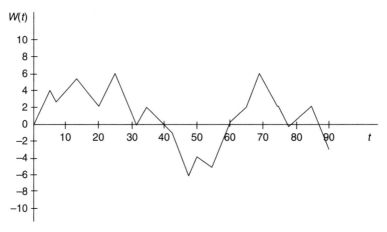

FIGURE 5.1 Sample Function of a Wiener Process.

A Brownian motion is sometimes called a *Wiener process*. A sample function of the Weiner process is shown in Figure 5.1.

Let $B(t) = W(t)/\sigma$. Then $E[B(t)] = 0$ and $\sigma^2_{B(t)} = 1$. The stochastic process $\{B(t), t \geq 0\}$ is called the *standard Brownian motion*, which has the property that when sampled at regular intervals it produces a symmetric random walk. Note that $B(t) \sim N(0, t)$. In the remainder of this chapter, we refer to the Weiner process $W(t) \sim N(0, \sigma^2 t)$ as the *classical Brownian motion* and use the two terms interchangeably.

5.2 BROWNIAN MOTION WITH DRIFT

Brownian motion is used to model stock prices. Because stock prices do not generally have a zero mean, it is customary to include a *drift* measure that makes the following model with a *drift rate* $\mu > 0$ a better model than the classical Brownian motion:

$$Y(t) = \mu t + W(t), \quad t \geq 0 \tag{5.2}$$

where $W(t)$ is the classical Brownian motion. Note that $E[Y(t)] = \mu t$ and $\sigma^2_{Y(t)} = \sigma^2 t$, which means that $Y(t) \sim N(\mu t, \sigma^2 t)$. Note also that we can express $Y(t)$ in terms of the standard Brownian motion as follows:

$$Y(t) = \mu t + \sigma B(t) \quad t \geq 0 \tag{5.3}$$

5.3 BROWNIAN MOTION AS A MARKOV PROCESS

Let $W(t)$ be a classical Brownian motion, and let $\mathfrak{I}_s$ denote the information that is being revealed by watching the process up through the time $s < t$. Then the conditional expected value of $W(t)$ given $\mathfrak{I}_s$ can be obtained as follows:

$$E[W(t) | \mathfrak{I}_s] = E[W(s) | \mathfrak{I}_s] + E\left[\{W(t) - W(s)\} | \mathfrak{I}_s\right]$$

The first term on the right-hand side is equal to $W(s)$ because it is already revealed through $\Im_s$. Also, the increment $W(t) - W(s)$ is independent of $\Im_s$; thus, $E[\{W(t) - W(s)\} | \Im_s] = E[W(t) - W(s)]$, which is 0. Therefore, we obtain

$$E[W(t) | \Im_s] = W(s) + E[W(t) | W(s)]$$

This means that to predict $W(t)$ given all the information up through time s, we only need to consider the value of the process at time s. Thus, a Brownian motion is a Markov process. This is not surprising because it is an independent increment process. As stated earlier, all independent increment processes have the Markov property.

Because for the classical Brownian motion the increment over an interval of length X has the Gaussian distribution with the probability density function (PDF)

$$f_X(x) = \frac{1}{\sqrt{2\pi\sigma^2}} e^{-x^2/2\sigma^2} \tag{5.4}$$

the classical Brownian motion is a Markov process with transition PDF given by

$$f_{Y|X}(y|x) = \frac{1}{\sqrt{2\pi\sigma^2}} e^{-(y-x)^2/2\sigma^2} \tag{5.5}$$

5.4 BROWNIAN MOTION AS A MARTINGALE

Let $0 \le s \le t$ and let $v \ge 0$. We show that $E[W(t+v)|W(s)] = W(t)$. We recall the Markov property that for $0 \le s \le t$, $E[W(t+v)|W(s)] = E[W(t+v)|W(t)]$. Therefore,

$$\begin{aligned}
E[W(t+v) | W(s)] &= E[W(t+v) | W(t)] = E[W(t) + \{W(t+v) - W(t)\} | W(t)] \\
&= W(t) + E[\{W(t+v) - W(t)\} | W(t)] \\
&= W(t) + E[W(t+v) - W(t)] = W(t) + E[W(v)] = W(t) + 0 \\
&= W(t)
\end{aligned}$$

where the fifth equality follows from the independent increments property.

5.5 FIRST PASSAGE TIME OF A BROWNIAN MOTION

Let T_k denote the time it takes a classical Brownian motion to go from $W(0) = 0$ to $W(t) = k \ne 0$; that is,

$$T_k = \min\{t > 0 : W(t) = k\} \tag{5.6}$$

Suppose $k > 0$. Then we obtain the PDF of T_k as follows:

$$P[W(t) \geq k] = P[W(t) \geq k \mid T_k \leq t]P[T_k \leq t] + P[W(t) \geq k \mid T_k > t]P[T_k > t]$$

By the definition of the first passage time, $P[W(t) \geq k \mid T_k > t] = 0$ for all t. Also, if $T_k \leq t$, then we assume that there exists a $t_0 \in (0, t)$ with the property that $W(t_0) = k$. We know that the process $\{W(t) \mid W(t_0) = k\}$ has a normal distribution with mean k and variance $\sigma^2(t - t_0)$. That is, the random variable

$$\{W(t) \mid W(t_0) = k\} \sim N(k, \sigma^2(t - t_0)) \quad \text{for all } t \geq t_0$$

Thus, from the symmetry about k, we have that

$$P[W(t) \geq k \mid T_k \leq t] = \frac{1}{2}$$

which means that the cumulative distribution function (CDF) of T_k is given by

$$F_{T_k}(t) = P[T_k \leq t] = 2P[W(t) \geq k] = \frac{2}{\sigma\sqrt{2\pi t}} \int_k^\infty \exp\{-v^2/2\sigma^2 t\} dv$$

By symmetry, T_k and T_{-k} are identically distributed random variables, which means that

$$F_{T_k}(t) = \frac{2}{\sigma\sqrt{2\pi t}} \int_{|k|}^\infty \exp\{-v^2/2\sigma^2 t\} dv = 2\left\{1 - \Phi\left(\frac{|k|}{\sigma\sqrt{t}}\right)\right\} \quad t > 0$$

where $\Phi(\cdot)$ is the CDF of the standard normal random variable. Let $y^2 = v^2/\sigma^2 t$, which means that $dv = 2dy\sqrt{t}$. Then the CDF becomes

$$F_{T_k}(t) = \frac{2}{\sqrt{2\pi}} \int_{|k|/\sigma\sqrt{t}}^\infty \exp\{-y^2/2\} dy \quad t > 0 \tag{5.7}$$

Differentiating with respect to t we obtain the PDF of T_k as follows:

$$f_{T_k}(t) = \frac{|k|}{\sqrt{2\pi\sigma^2 t^3}} \exp\left\{-\frac{k^2}{2\sigma^2 t}\right\} \quad t > 0 \tag{5.8}$$

Because for the standard Brownian motion $\sigma = 1$, the PDF of T_k for the standard Brownian motion is given by

$$f_{T_k}(t) = \frac{|k|}{\sqrt{2\pi t^3}} \exp\left\{-\frac{k^2}{2t}\right\} \quad t > 0 \tag{5.9}$$

5.6   MAXIMUM OF A BROWNIAN MOTION

Let $M(t)$ denote the maximum value of the classical Brownian motion $\{W(t)\}$ in the interval $[0, t]$; that is,

$$M(t) = \max\{W(u), 0 \le u \le t\} \tag{5.10}$$

The PDF of $M(t)$ is obtained by noting that

$$P\big[M(t) \ge x\big] = P\big[T_x \le t\big]$$

Thus, we have that

$$F_{M(t)}(x) = P[M(t) \le x] = 1 - P[T_x \le t] = 1 - 2\left\{1 - \Phi\left(\frac{|k|}{\sigma\sqrt{t}}\right)\right\}$$

$$= 1 - \frac{2}{\sqrt{2\pi}} \int_{|x|/\sigma\sqrt{t}}^{\infty} \exp\{-y^2/2\}\, dy$$

Upon differentiation we obtain the PDF as

$$f_{M(t)}(x) = \sqrt{\frac{2}{\pi\sigma^2 t}} \exp\left\{-\frac{x^2}{2\sigma^2 t}\right\} \quad x \ge 0 \tag{5.11}$$

5.7   FIRST PASSAGE TIME IN AN INTERVAL

Let T_{ab} denote the time at which the classical Brownian motion $\{W(t), t \ge 0\}$ for the first time hits either the value a or the value b, where $b < 0 < a$, as shown in Figure 5.2. Thus, we can write

$$T_{ab} = \min\{t : W(t) = a \quad \text{or} \quad W(t) = b\} \quad b < 0 < a < \infty$$

Let p_{ab} be the probability that $\{W(t)\}$ assumes the value a first; that is,

$$p_{ab} = P[W(T_{ab}) = a]$$

Now, T_{ab} is a stopping time whose mean $E[T_{ab}]$ is finite. Thus, according to the stopping time theorem, $E[T_{ab}] = E[W(0)] = 0$. This means that

$$E[W(T_{ab})] = a p_{ab} + b(1 - p_{ab}) = 0$$

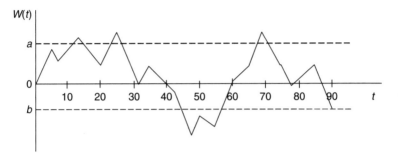

FIGURE 5.2   First Passage Time in an Interval.

From this we obtain the probability that the process hits a before b as

$$p_{ab} = \frac{|b|}{a + |b|} \tag{5.12}$$

An alternative method of solving the problem is by seeing the Brownian motion as a limit of the symmetric random walk. The probability that the process hits a before b can then be likened to the gambler's ruin problem in which the process is equally likely to go up or down by a distance of Δh. Thus, with respect to the gambler's ruin problem of Chapter 3, $N = (a + |b|)/\Delta h$ and $i = |b|/\Delta h$. Because p_{ab} is the ruin probability, and for the symmetric walk the ruin probability is i/N, we have that

$$p_{ab} = \frac{i}{N} = \frac{|b|/\Delta h}{(a + |b|)/\Delta h} = \frac{|b|}{a \, | \, |b|}$$

5.8   THE BROWNIAN BRIDGE

The Brownian bridge is a classical Brownian motion defined on the interval $[0,1]$ and conditioned on the event $W(1) = 0$. Thus, the Brownian bridge is the process $\{W(t), \, t \in [0,1] | W(1) = 0\}$. One way to realize the process is by defining $X(t)$, the Brownian bridge, as follows:

$$X(t) = W(t) - tW(1) \quad 0 \le t \le 1 \tag{5.13}$$

The Brownian bridge is sometimes called the *tied-down Brownian motion* (or *tied-down Wiener process*). It is useful for modeling a system that starts at some given level and is expected to return to that level at some specified future time. We note that

$$X(0) = W(0) - 0W(1) = 0$$
$$X(1) = W(1) - 1W(1) = 0$$

Thus, $E[X(t)] = 0$. For $0 \le s < t \le 1$, the covariance of $X(t)$ and $X(s)$ is given by

$$\begin{aligned}
\text{Cov}\{X(s)X(t)\} &= E[\{X(s)-E[X(s)]\}\{X(t)-E[X(t)]\}] = E[X(s)X(t)] \\
&= E[\{W(s)-sW(1)\}\{W(t)-tW(1)\}] \\
&= E[W(s)W(t)-tW(s)W(1)-sW(t)W(1)+stW^2(1)] \\
&= \sigma^2(s\wedge t)-\sigma^2 t(s\wedge 1)-\sigma^2 s(t\wedge 1)+\sigma^2 st \\
&= \sigma^2\{s-st-st+st\} = \sigma^2(s-st) \\
&= \sigma^2 s(1-t) \tag{5.14}
\end{aligned}$$

where we have used the fact that $E[W(s)W(t)]=\sigma^2\min(s, t)=\sigma^2(s\wedge t)$. Thus, the Brownian bridge is not a wide-sense stationary process because the covariance $\text{Cov}\{X(s)X(t)\}$ is not a function of only the difference between s and t.

5.9   GEOMETRIC BROWNIAN MOTION

Let $\{X(t), t\geq 0\}$ be a Brownian motion with drift. The process $\{Y(t), t\geq 0\}$, which is defined by

$$Y(t) = e^{X(t)} \tag{5.15}$$

is called the geometric Brownian motion. We will discuss the process in greater details after our discussion on stochastic differential equations in Chapter 6.

5.10   THE LANGEVIN EQUATION

The Langevin equation is the most widely known mathematical model of Brownian motion. Langevin wrote down the equation of motion of the Brownian particle according to Newton's laws with the assumption that the particle experiences two forces:

1. A viscous drag of magnitude $-\xi(dx(t)/dt)=-\xi\dot{x}(t)$, which represents a dynamical friction experienced by the particle, where $x(t)$ is the displacement at time t and ξ is the coefficient of friction
2. A rapidly fluctuating force $F(t)$ that is due to the impacts of the molecules of the fluid on the particle; this is a zero-mean white noise.

We consider only one particle and proceed to calculate the mean-square displacement (MSD) for the particle at time t. If $x(t)$ is the location of a particles at time t, the MSD is defined by

$$\text{MSD}(t) = E[\{x(t)-x(0)\}^2]$$

Thus, if we assume that there are N particles, then

$$\text{MSD}(t) = \frac{1}{N} \sum_{k=1}^{N} E[\{x_k(t) - x_k(0)\}^2]$$

If m is the mass of the particle, then according to Newton's second law, the equation of motion is given by

$$m\frac{d^2x(t)}{dt^2} = -\xi\frac{dx(t)}{dt} + F(t) \tag{5.16}$$

The friction term $\xi(dx(t)/dt)$ obeys the Stokes' law, which states that the frictional force decelerating a spherical particle of radius r is

$$\xi\frac{dx(t)}{dt} = 6\pi\eta r\frac{dx(t)}{dt} \tag{5.17}$$

where η is the viscosity of the surrounding fluid. With respect to $F(t)$, it is assumed that it is independent of $x(t)$ and $F(t)$ varies extremely rapidly compared to $x(t)$ with mean $E[F(t)] = 0$.

Multiplying Equation 5.16 by $x(t)$ we obtain

$$mx(t)\frac{d^2x(t)}{dt^2} = -\xi x(t)\frac{dx(t)}{dt} + F(t)x(t) \tag{5.18}$$

Now,

$$x(t)\frac{dx(t)}{dt} = \frac{1}{2}\frac{dx^2(t)}{dt}$$

$$x(t)\frac{d^2x(t)}{dt^2} = \frac{1}{2}\frac{d}{dt}\left\{\frac{dx^2(t)}{dt}\right\} - \left\{\frac{dx(t)}{dt}\right\}^2$$

Thus, Equation 5.18 becomes

$$\frac{m}{2}\frac{d}{dt}\left\{\frac{dx^2(t)}{dt}\right\} - m\left\{\frac{dx(t)}{dt}\right\}^2 = -\frac{\xi}{2}\frac{dx^2(t)}{dt} + F(t)x(t) \tag{5.19}$$

The preceding equation applies to only one particle. Thus, to account for the macroscopic system we average the equation to get

$$\frac{m}{2}\frac{d}{dt}\left\{\frac{dE[x^2(t)]}{dt}\right\} - mE\left[\left\{\frac{dx(t)}{dt}\right\}^2\right] = -\frac{\xi}{2}\frac{dE[x^2(t)]}{dt} + E[F(t)x(t)] \tag{5.20}$$

Since we assume that $F(t)$ and $x(t)$ are independent, and since $E[F(t)]=0$, we have that $E[F(t)x(t)]=E[F(t)]E[x(t)]=0$. Also, the *equipartition theorem* states that energy is shared equally among all energetically accessible degrees of freedom of a system (Huang 1987). This means that when the velocity process has reached its equilibrium value, the Maxwellian distribution can be assumed to hold so that the mean kinetic energy $(1/2)m\overline{v^2}$ of the Brownian particle of mass m and velocity v reaches a value $(1/2)kT$; that is,

$$\frac{1}{2}mE\left[\left\{\frac{dx(t)}{dt}\right\}^2\right]=\frac{1}{2}kT$$

where k is the Boltzmann constant and T is the absolute temperature. Thus, Equation 5.20 becomes

$$\frac{m}{2}\frac{d}{dt}\left\{\frac{dE[x^2(t)]}{dt}\right\}+\frac{\xi}{2}\frac{dE[x^2(t)]}{dt}=kT \tag{5.21}$$

Define

$$y=\frac{dE[x^2(t)]}{dt}$$

Then, Equation 5.21 becomes

$$\frac{dy}{dt}+\frac{\xi}{m}y=\frac{dy}{dt}+\frac{y}{\tau}=\frac{2kT}{m} \tag{5.22}$$

where $\tau=m/\xi$. The solution to Equation 5.22 is

$$ye^{x/\tau}=\int\frac{2kT}{m}e^{x/\tau}dx$$

which gives

$$y=\frac{dE[x^2(t)]}{dt}=2E\left[x(t)\frac{dx(t)}{dt}\right]=Ce^{-\xi t/m}+\frac{2kT}{\xi} \tag{5.23}$$

where C is a constant of integration. Let $\gamma=\xi/m$ so that the Equation 5.23 becomes

$$E\left[x(t)\frac{dx(t)}{dt}\right]=Ce^{-\gamma t}+\frac{2kT}{\xi} \tag{5.24}$$

If we assume the initial conditions $x(0) = 0$, then we have that $C = -2kT/\xi$. Thus, we have that

$$E\left[x(t)\frac{dx(t)}{dt}\right] = \frac{2kT}{\xi}(1-e^{-\gamma t}) \qquad (5.25)$$

Integrating with respect to t we obtain

$$\int_0^t E\left[x(u)\frac{dx(u)}{du}\right]du = \frac{1}{2}\int_0^t \frac{d}{du}E[x^2(u)]du = \frac{2kT}{\xi}\int_0^t (1-e^{-\gamma u})du$$

From this we obtain

$$E[x^2(t)] = \frac{4kT}{\xi}\left\{t - \frac{1}{\gamma}(1-e^{-\gamma t})\right\} \qquad (5.26)$$

This is the MSD of a Brownian particle at time t. We consider two cases:

1. When $t << \gamma^{-1}$, we can use the Taylor series to write $e^{-\gamma t} = 1 - \gamma t + \frac{1}{2}\gamma^2 t^2 - \cdots$ and obtain

$$\begin{aligned} E[x^2(t)] &= \frac{4kT}{\xi}\left\{t - \frac{1}{\gamma}(1-e^{-\gamma t})\right\} = \frac{2kT}{\xi}\left\{t - \frac{1}{\gamma}\left(1 - 1 + \gamma t - \frac{1}{2}\gamma^2 t^2 + \cdots\right)\right\} \\ &= \frac{4kT}{\xi}\left\{t - t + \frac{1}{2}\gamma t^2 - \cdots\right\} \\ &= \frac{2kT}{\xi}t^2 \end{aligned} \qquad (5.27)$$

2. When $t >> \gamma^{-1}$, $e^{-\gamma t} \to 0$. Thus, Equation 5.26 becomes

$$E[x^2(t)] = \frac{4kT}{\xi}t = Dt \Rightarrow E[x^2(t)] \propto t \qquad (5.28)$$

This is the formula that Einstein obtained as the formula for the MSD. Under this condition the particle exhibits diffusive movement, and D is defined as the diffusion coefficient.

5.11 SUMMARY

This chapter has discussed the basics of the Brownian motion by introducing the essential properties of the process. Other aspects of the process will be discussed in Chapter 6 following the introduction to stochastic calculus. As discussed earlier, the Brownian motion is the central theme of stochastic calculus. Finally, the Langevin equation, which is the most widely known mathematical model of Brownian motion, is used to obtain the MSD of the process.

PROBLEMS

5.1 Assume that X and Y are independent random variables such that $X \sim N(0, \sigma^2)$ and $Y \sim N(0, \sigma^2)$. Consider the random variables $U = (X+Y)/2$ and $V = (X-Y)/2$. Show that U and V are independent with $U \sim N(0, \sigma^2/2)$ and $V \sim N(0, \sigma^2/2)$.

5.2 Suppose $X(t)$ is a standard Brownian motion and $Y(t) = tX(1/t)$. Show that $Y(t)$ is a standard Brownian motion.

5.3 Let $\{X(t), t \geq 0\}$ be a Brownian motion with drift rate μ and variance parameter σ^2. What is the conditional distribution of $X(t)$ given that $X(u) = b$, $u < t$?

5.4 Let $T = \min\{t | B(t) = 5 - 3t\}$. Use the martingale stopping theorem to find $E[T]$.

5.5 Let $Y(t) = \int_0^t B(u)du$, where $\{B(t), t \geq 0\}$ is the standard Brownian motion. Find

1. $E[Y(t)]$
2. $E[Y^2(t)]$
3. The conditional distribution of $Y(t)$, given that $B(t) = x$.

5.6 Consider the Brownian motion with drift

$Y(t) = \mu t + \sigma B(t) + x$

where $Y(0) = x$ and $b < x < a$. Let $p_a(x)$ denote the probability that hits a before b.
1. Show that

$$\frac{1}{2} \frac{d^2 p_a(x)}{dx^2} + \mu \frac{dp_a(x)}{dx} = 0$$

2. Deduce that

$$p_a(x) = \frac{e^{-2\mu b} - e^{-2\mu x}}{e^{-2\mu b} - e^{-2\mu a}}$$

3. What is $p_a(x)$ when $\mu = 0$?

5.7   What is the mean value of the first passage time of the reflected Brownian motion $\{|B(t)|, t \geq 0\}$ with respect to a positive level x, where $B(t)$ is the standard Brownian motion? Determine the CDF of $|B(t)|$.

5.8   Let the process $\{X(t), t \geq 0\}$ be defined by $X(t) = B^2(t) - t$ where $\{B(t), t \geq 0\}$ is a standard Brownian motion.

   1. What is $E[X(t)]$?

   2. Show that $\{X(t), t \geq 0\}$ is a martingale.

   Hint: Start by computing $E[X(t)|B(v), 0 \leq v \leq t]$.

CHAPTER 6

INTRODUCTION TO STOCHASTIC CALCULUS

6.1 INTRODUCTION

One of the basic tools for analyzing the Brownian motion is stochastic calculus, which deals with infinitesimal calculus on nondifferentiable functions. It arises from the need to incorporate unpredictable random factors in system modeling. The primary focus of stochastic calculus is the Brownian motion because it is a model that is not only useful and permits explicit calculations to be performed, but also it is applicable to many naturally occurring phenomena.

To motivate the discussion, consider a system whose state at time t can be represented by $X(t)$, and let the initial value be $X(0) = x_0$. The dynamics of the system can be described by the ordinary differential equation (ODE):

$$\frac{dX(t)}{dt} = a(X(t)), \quad t \geq 0, \ X(0) = x_0 \tag{6.1}$$

where $a(\cdot)$ is a given smooth function. The solution to this equation is the trajectory of the system, which is

$$X(t) = x_0 + \int_0^t a\big(X(u)\big) du$$

Elements of Random Walk and Diffusion Processes, First Edition. Oliver C. Ibe.
© 2013 John Wiley & Sons, Inc. Published 2013 by John Wiley & Sons, Inc.

This is the so-called Riemann integral, which is a part of deterministic calculus that deals with differentiation and integration of deterministic functions. Unfortunately in many applications, the experimentally measured trajectories of systems modeled by the preceding ODE do not behave as predicted due to some random effects that disturb the system. This implies that the ODE needs to be modified to include these random effects. This goal is achieved by rewriting the system equation as follows:

$$\frac{dX(t)}{dt} = a\big(X(t)\big) + b\big(X(t)\big)\xi(t), \quad t \geq 0, X(0) = x_0 \tag{6.2}$$

where $\xi(t)$ represents a white noise and $b(\cdot)$ is a given function. The common practice is to represent this white noise by the time derivative of the Wiener process (or Brownian motion). That is,

$$\xi(t) = \frac{dW(t)}{dt}$$

Thus, Equation 6.2 becomes

$$\frac{dX(t)}{dt} = a\big(X(t)\big) + b\big(X(t)\big)\frac{dW(t)}{dt}, \quad X(0) = x_0 \tag{6.3}$$

When we multiply the equation by "dt," we obtain

$$dX(t) = a\big(X(t)\big)dt + b\big(X(t)\big)dW(t), \quad X(0) = x_0 \tag{6.4}$$

We know that the path of Brownian motion takes sharp turns everywhere and thus Brownian motion is nondifferentiable. Stochastic calculus deals with integration of a stochastic process with respect to another stochastic process. Because Brownian motion is nowhere differentiable, any stochastic process that is driven by Brownian motion is nowhere differentiable. The driving force behind stochastic calculus was the attempt to understand the motion driven by a series of small random impulses, which are modeled by Brownian motion. Because the net distance traveled in time Δt by a particle in Brownian motion is proportional to $\sqrt{\Delta t}$, in the stochastic differential equation (SDE) that represents the dynamic behavior of the process, future changes are expressed as differentials, not as derivatives. For a process $\{X(t), t \geq 0\}$, the differential $dX(t)$ is defined by

$$dX(t) = X(t+dt) - X(t) \tag{6.5}$$

The integral form is the forward sum of uncountable and random increments over time and is given by

$$X(t) = \int_0^t dX(u)$$

Stochastic calculus is the mathematics used for modeling financial options. It is used to model investor behavior and asset pricing. It has also found applications in such fields as control theory and mathematical biology. Observe that $X(t)$ is a random variable, and we would like to obtain such statistics as its mean and variance.

6.2   THE ITO INTEGRAL

The Ito integral deals with the integration of expressions of the form

$$\int_0^t X(u)dW(u)$$

where $X(t)$ is a stochastic process and $W(t)$ is the classical Brownian motion. Thus, the earlier integral is referred to as the Ito integral of $X(t)$ with respect to the Brownian motion. If $W(t)$ were the function $w(t)$ that is differentiable, we would write $dw(u)=w_1(u)du$, where $w_1(u)=dw(u)/du$, and thus obtain the expression

$$\int_0^t X(u)\,dw(u) = \int_0^t X(u)w_1(u)du$$

that is known to be a standard integral. When the process $W(t)$ is not differentiable, as is the case when it is the Brownian motion, the integral becomes an unfamiliar integral. To evaluate the integral when $W(t)$ is not differentiable we first divide the interval $[0, t]$ into n disjoint intervals at points $0=t_0<t_1<\cdots<t_n=t$ and obtain

$$\int_0^t X(u)dw(u) = \lim_{n\to\infty} \sum_{k=0}^{n-1} X(u_k)\left\{W(u_{k+1})-W(u_k)\right\} \qquad (6.6)$$

which is the *Ito integral*. However, for this to work $X(t)$ and $W(t)$ are required to satisfy certain conditions, which include the fact that $X(t)$ must be smooth enough for $X(t_k)$ to represent $X(t)$ in the interval (t_k, t_{k+1}). Also, the $X(t_k)$ are required to be independent of the increments $W(t_{k+1})-W(t_k)$. In addition, $X(t)$ is required to be an *adapted process*, which is sometimes called a *nonanticipating process* because it cannot "see into the future."

To fully understand an adapted process, we first define the concept of *filtration*. A filtration is a family of σ-algebras $\{F_t, 0\le t<\infty\}$ that is increasing; that is, if $s\le t$, then $F_s\subset F_t$. The property that a filtration is increasing implies that information is not forgotten. The space $(\Omega, F, \{F_t\}, P)$ is called a *filtered probability space*. A stochastic process $\{X(t), 0\le t\le T\}$ on a filtered probability space is called an adapted process if for any $t\in [0, T]$, F_t contains all the information about the random variable $X(t)$; alternatively, it is an adapted process if the random variable $X(t)$ is F_t-adapted.

Thus, a stochastic process $\{X(t)\}$ is defined to be *Ito integrable* on the interval $[0, T]$ if the following conditions are satisfied:

1. The random variable $X(t)$ is adapted for $t \in [0, T]$.

2. $\int_0^t E\left[X^2(u)\right] du < \infty.$

We state the following propositions without proof; the proofs can be found in many books on stochastic processes such as Capasso and Bakstein (2005), Steele (2001), Oksendal (2005), Benth (2004), and Klebaner (2005).

Proposition 6.1: Let $f(t)$ and $g(t)$ be continuous functions in the interval $[a, b]$, and let $B(t)$ be a standard Brownian motion. Then

1. $E\left[\int_a^b f(t)dB(t)\right] = 0$

2. $E\left[\int_a^b f(t)\,dB(t)\int_a^b g(t)\,dB(t)\right] = \int_a^b E\left[f(t)g(t)\right] dt$

3. $E\left[\left\{\int_a^b f(t)\,dB(t)\right\}^2\right] = \mathrm{Var}\left(\int_a^b f(t)\,dB(t)\right) = \int_a^b E\left[\{f(t)\}^2\right] dt;$ this is called the

   Ito isometry property.

Proposition 6.2:

$$\int_a^b B(t)\,dB(t) = \frac{1}{2}\left[B^2(b) - B^2(a)\right] - \frac{b-a}{2} = \frac{B^2(b) - B^2(a) - (b-a)}{2}.$$

Proposition 6.3:

$$\int_a^b t\,dB(t) = bB(b) - aB(a) - \int_a^b B(t)dt$$

6.3 THE STOCHASTIC DIFFERENTIAL

Let $\{X(t),\ t \in [0, T]\}$ be a stochastic process such that for $0 < t \le T$ we have the following stochastic integral:

$$X(t) = X(0) + \int_0^t a(u)\,du + \int_0^t b(u)dB(u) \tag{6.7}$$

where $a(t)$ and $b(t)$ are continuous functions in the interval $[0, T]$. Then we say that $X(t)$ has the stochastic differential

$$dX(t) = a(t)dt + b(t)dB(t) \tag{6.8}$$

in the interval $[0, T]$. Observe that the stochastic integral involves both the Ito and Riemann integrals of an unknown stochastic process. The first integral in Equation 6.7 is a Riemann integral while the second integral is an Ito integral. Similarly, the stochastic differential of Equation 6.8 is the sum of two differentials that are associated with the Riemann integral and the Ito integral.

6.4 THE ITO'S FORMULA

The Ito's formula serves as a bridge between classical theory and stochastic theory. It is the stochastic equivalent of Taylor's theorem about the expansion of functions. Consider the stochastic process $X(t)$ with the stochastic differential

$$dX(t) = a(t)dt + b(t)dB(t)$$

in the interval $[0, T]$. Let $f(t, y)$ be a continuous function in the interval $[0, T]$ that is twice differentiable in y. Then the Ito's formula gives the stochastic differential of the function $y(t) = f(t, X(t))$ as follows:

$$dy(t) = \left\{ \frac{\partial f(X(t), t)}{\partial t} + a(t)\frac{\partial f(X(t), t)}{\partial X} + \frac{1}{2}b^2(t)\frac{\partial^2 f(X(t), t)}{\partial X^2} \right\} dt$$
$$+ b(t)\frac{\partial f(X(t), t)}{\partial X} dB(t) \tag{6.9}$$

Note that the first part of the formula is the consequence of the Taylor formula. The last term is essentially a new term in the "stochastic Taylor series." Also, the Ito's formula can be viewed as the stochastic analog of the chain rule of classical calculus.

6.5 STOCHASTIC DIFFERENTIAL EQUATIONS

A differential equation is a rule that allows us to calculate the value of some quantity at a later time, given the value at some earlier time. Thus, an SDE can be viewed as the representation of the dynamic behavior of a stochastic process. The dynamic behavior of almost all important continuous stochastic processes can be expressed by an equation of the form

$$dX(t) = \mu\big(X(t), t\big)dt + \sigma\big(X(t), t\big)dB(t) \quad X(0) = x_0 \tag{6.10}$$

where the functions $\mu(X(t), t)$ and $\sigma(X(t), t)$ are given, $X(t)$ is an unknown process, and $B(t)$ is the standard Brownian motion. Such an SDE is said to be driven by Brownian motion. As stated earlier, Brownian motion-driven systems have become an important framework for modeling financial markets. The coefficients $\mu(X(t), t)$ and $\sigma(X(t), t)$ can be interpreted as measures of short-term growth and short-term variability, respectively. Thus, adjusting them permits a modeler to construct stochastic processes that reflect real-life behavior. A solution to the preceding SDE is

$$X(t) = X(0) + \int_0^t \mu\big(X(u), u\big)du + \int_0^t \sigma\big(X(u), u\big)dB(u) \tag{6.11}$$

If the two integrals exist for all $t>0$, $X(t)$ is called a strong solution of the SDE. In some processes, the coefficients $\mu(X(t), t)$ and $\sigma(X(t), t)$ are defined as follows: $\mu(X(t), t) = \mu X(t)$ where $-\infty < \mu < \infty$, and $\sigma(X(t), t) = \sigma X(t)$ where $\sigma > 0$. Thus, the SDE for such processes is given by

$$dX(t) = \mu X(t)dt + \sigma X(t)dB(t) \quad X(0) = x_0 > 0 \tag{6.12}$$

The Ito's formula is the key to the solution of many SDEs. We illustrate the solution with the geometric Brownian motion.

Example 6.1: We are given $dX(t) = \mu dt + \sigma dB(t)$ with $X(0) = x_0$. The solution is

$$X(t) = X(0) + \int_0^t \mu du + \int_0^t \sigma dB(u) = X(0) + \mu t + \sigma B(t)$$

Example 6.2: We are given $dX(t) = \mu dt + \sigma t dB(t)$ with $X(0) = x_0$. The solution is

$$X(t) = X(0) + \int_0^t \mu du + \int_0^t \sigma u dB(u)$$

$$= X(0) + \mu t + \sigma t B(t) - \sigma 0 B(0) - \int_0^t B(u)du = X(0) + \mu t + \sigma t B(t) - \int_0^t B(u)du$$

6.6   SOLUTION OF THE GEOMETRIC BROWNIAN MOTION

As discussed in Chapter 5, we can define the geometric Brownian motion $\{Y(t), t \geq 0\}$ as follows:

$$Y(t) = e^{X(t)}$$

where $\{X(t),\ t \geq 0\}$ is a Brownian motion with drift. The SDE of the process is obtained as follows. Because $\ln\{Y(t)\} = X(t)$, we have that

$$\frac{dY(t)}{Y(t)} = dX(t) = \mu dt + \sigma dB(t) \tag{6.13}$$

The equation can also be written as follows:

$$dY(t) = \mu Y(t)\,dt + \sigma Y(t)\,dB(t) \tag{6.14}$$

The solution to the SDE of the process is obtained by defining $f(Y(t)) = \ln\{Y(t)\} = X(t)$, so that

$$\frac{\partial f}{\partial t} = 0, \quad \frac{\partial f}{\partial Y} = \frac{1}{Y}, \quad \frac{\partial^2 f}{\partial Y^2} = -\frac{1}{Y^2}$$

Since $a(t) = \mu Y(t)$ and $b(t) = \sigma Y(t)$, then from the Ito's formula we have that

$$df(Y(t),t) = d\left[\ln\{Y(t)\}\right] = \left[(0) + \frac{\mu Y(t)}{Y(t)} + \frac{1}{2}\left\{-\frac{1}{Y^2(t)}\right\}\sigma^2 Y^2(t)\right]dt + \frac{\sigma Y(t)}{Y(t)}dB(t)$$

$$= \left\{\mu - \frac{1}{2}\sigma^2\right\}dt + \sigma dB(t)$$

Now we know that

$$d\left[\ln\{Y(t)\}\right] = \frac{dY(t)}{Y(t)} = \left\{\mu - \frac{1}{2}\sigma^2\right\}dt + \sigma dB(t)$$

From this we obtain

$$\int_0^t d\ln\{Y(u)\} = \int_0^t \left\{\mu - \frac{1}{2}\sigma^2\right\}du + \int_0^t \sigma dB(u)$$

Thus, since $B(0) = 0$,

$$\ln\left\{\frac{Y(t)}{Y(0)}\right\} = \left\{\mu - \frac{1}{2}\sigma^2\right\}t + \sigma B(t) \tag{6.15}$$

Finally, we get

$$Y(t) = Y(0)\exp\left\{\left[\mu - \frac{1}{2}\sigma^2\right]t + \sigma B(t)\right\}$$

$$= Y(0)\exp\left\{\left[\mu - \frac{1}{2}\sigma^2\right]t + W(t)\right\} \tag{6.16}$$

where $W(t)$ is the classical Brownian motion. Thus, we may express the solution to the equation as $Y(t) = Y_0 e^{X(t)}$, where

$$X(t) = \left[\mu - \frac{1}{2}\sigma^2\right]t + \sigma B(t) \Rightarrow dX(t) = \left[\mu - \frac{1}{2}\sigma^2\right]dt + \sigma dB(t)$$

This means that $X(t)$ is a Brownian motion with drift.

Example 6.3: Stock X follows a geometric Brownian motion where the drift factor is 0.96 and the variance factor is 0.55. At some particular time t, it is known that $dt = 0.04$, and $dB(t) = 0.45$. At time t, the stock trades for $200 per share. What is the instantaneous rate of change in the price of stock X?

Solution: The instantaneous rate of change in the price of the stock is given by

$$dX(t) = \mu X(t)dt + \sigma X(t)dB(t)$$

where $X(t) = 200$, $\mu = 0.96$ and $\sigma = 0.55$. Thus, we have that

$$dX(t) = (0.96)(200)(0.04) + (0.55)(200)(0.45) = 57.18$$

Example 6.4: A stock X follows a geometric Brownian motion with a drift factor of 0.35 and a volatility of 0.43. Given that $X(4) = 2$, what is the probability that $X(13) > 9$?

Solution: We are required to find $P[X(13) > 9 | X(4) = 2]$. First, we use the formula

$$X(t) = X(u)\exp\left\{\left[\mu - \frac{1}{2}\sigma^2\right](t-u) + \sigma\left(\sqrt{t-u}\right)B(t)\right\}$$

Thus,

$$X(13) = X(4)\exp\left\{[0.35 - (0.5)(0.43)^2](13-4) + 0.43\left(\sqrt{13-4}\right)B(13)\right\}$$

$$= 2\exp\left\{2.31795 + 1.29 B(13)\right\}$$

From this we obtain

$$
\begin{aligned}
P[X(13) > 9 \mid X(4) = 2] &= P[2\exp\{2.31795 + 1.29B(13)\} > 9] \\
&= P[\exp\{2.31795 + 1.29B(13)\} > 4.5] \\
&= P[2.31795 + 1.29B(13) > \ln(4.5)] \\
&= P[2.31795 + 1.29B(13) > 1.504077397] \\
&= P[1.29B(13) > -0.8138726] = P[B(13) > -0.63091] \\
&= 1 - P[B(13) \le -0.63091] = 1 - \Phi(-0.63091) \\
&= 1 - \left\{ 1 - \Phi(0.63091) \right\} = \Phi(0.63091) \approx \Phi(0.63) \\
&= 0.7357
\end{aligned}
$$

where we have used the fact that $B(t) \sim N(0,1)$.

6.7 THE ORNSTEIN–UHLENBECK PROCESS

The Brownian motion is used to construct the Ornstein–Uhlenbeck (OU) process, which has become a popular tool for modeling interest rates. Recall that the derivative of the Brownian motion $X(t)$ does not exist at any point in time. Thus, if $X(t)$ represents the position of a particle, we might be interested in obtaining its velocity, which is the derivative of the motion. The OU process is an alternative model of the Brownian motion that overcomes the preceding problem. It does this by considering the velocity $V(t)$ of a Brownian motion at time t. Over a small time interval, two factors affect the change in velocity: the frictional resistance of the surrounding medium whose effect is proportional to $V(t)$, and the random impact of neighboring particles whose effect can be represented by a standard Wiener process. Thus, because mass times velocity equals force, we have that

$$
mdV(t) = -\gamma V(t)dt + dB(t)
$$

where $\gamma > 0$ is called the *friction coefficient* and $m > 0$ is the mass. If we define $\alpha = \gamma/m$ and $\beta = 1/m$ we obtain the OU process with the following differential equation:

$$
dV(t) = -\alpha V(t)dt + \beta dB(t) \tag{6.17}
$$

The OU process is used to describe the velocity of a particle in a fluid and is encountered in statistical mechanics. It is the model of choice for random movement toward a concentration point. It is sometimes called a *continuous-time Gauss–Markov process*, where a Gauss–Markov process is a stochastic process that satisfies the requirements for both a Gaussian process and a Markov process. Because a Wiener process is both a Gaussian process and a Markov process, in addition to being a stationary independent increment process, it can be considered a Gauss–Markov process with independent increments.

The OU process can be obtained from the standard Brownian process $B(t)$ by scaling and time change, as follows:

$$V(t) = e^{-\alpha t} B\left(\frac{\beta^2}{2\alpha} e^{2\alpha t}\right) \tag{6.18}$$

Thus, $E[V(t)] = 0$, and because $B(t)$ is Gaussian, its covariance is

$$
\begin{aligned}
\text{Cov}\{V(t+\tau)V(t)\} &= E\left[e^{-\alpha(t+\tau)} B\left(\frac{\beta^2}{2\alpha} e^{2\alpha(t+\tau)}\right) e^{-\alpha t} B\left(\frac{\beta^2}{2\alpha} e^{2\alpha t}\right)\right] \\
&= e^{-\alpha(2t+\tau)} E\left[B\left(\frac{\beta^2}{2\alpha} e^{2\alpha(t+\tau)}\right) B\left(\frac{\beta^2}{2\alpha} e^{2\alpha t}\right)\right] \\
&= e^{-\alpha(2t+\tau)} \min\left\{\frac{\beta^2}{2\alpha} e^{2\alpha(t+\tau)}, \frac{\beta^2}{2\alpha} e^{2\alpha t}\right\} \\
&= e^{-\alpha(2t+\tau)} \frac{\beta^2}{2\alpha} e^{2\alpha t} = \frac{\beta^2}{2\alpha} e^{-\alpha\tau}
\end{aligned}
\tag{6.19}
$$

Thus, $V(t)$ is a stationary process. However, unlike the Wiener process, it does not have independent increments. For a *standard OU process*, $\alpha = \beta = 1$ so that

$$V(t) = e^{-t} B\left(\frac{1}{2} e^{2t}\right) \tag{6.20}$$

The aforementioned OU process is referred to as the *Brownian motion-driven OU process*. Other types of the process are the Poisson-driven and gamma-driven OU processes.

6.7.1 Solution of the Ornstein–Uhlenbeck SDE

To solve the OU SDE $dV(t) = -\alpha V(t)dt + \beta dB(t)$, we define $f(V(t), t) = e^{\alpha t} V(t)$. This gives

$$\frac{\partial f}{\partial t} = \alpha e^{\alpha t} V(t), \quad \frac{\partial f}{\partial V} = e^{\alpha t}, \quad \frac{\partial^2 f}{\partial V^2} = 0$$

Thus, applying the Ito's formula we obtain

$$
\begin{aligned}
df = d(e^{\alpha t} V(t)) &= \left\{\frac{\partial f}{\partial t} + (-\alpha V)\frac{\partial f}{\partial V} + \frac{1}{2}\beta^2 \frac{\partial^2 f}{\partial V^2}\right\} dt + \beta\frac{\partial f}{\partial V} dB \\
&= \left\{\alpha e^{\alpha t} V(t) - \alpha e^{\alpha t} V(t) + \frac{1}{2}\beta^2 (0)\right\} dt + \beta e^{\alpha t} dB \\
&= \beta e^{\alpha t} dB(t)
\end{aligned}
$$

Given that $V(0) = v_0$, we obtain the solution as

$$e^{\alpha t}V(t) = v_0 + \int_0^t e^{\alpha u}\beta dB(u)$$

which gives

$$V(t) = v_0 e^{-\alpha t} + e^{-\alpha t}\int_0^t e^{\alpha u}\beta dB(u)$$

Because $B(t)$ is a Brownian process, we have that the mean of $V(t)$ is

$$E[V(t)] = v_0 e^{-\alpha t} + E\left[\int_0^t e^{-\alpha(t-u)}\beta dB(u)\right] = v_0 e^{-\alpha t}$$

Using the Ito isometry we obtain the variance of $V(t)$ as

$$\sigma_{V(t)}^2 = E\left[\left\{\int_0^t e^{-\alpha(t-u)}\beta dB(u)\right\}^2\right] = \int_0^t \left[\beta e^{-\alpha(t-u)}\right]^2 du = \beta^2 \int_0^t e^{-2\alpha(t-u)}du$$

$$= \frac{\beta^2}{2\alpha}\left(1 - e^{-2\alpha t}\right)$$

Thus, we have that

$$V(t) \sim N\left(v_0 e^{-\alpha t},\ \frac{\beta^2}{2\alpha}\left[1 - e^{-2\alpha t}\right]\right) \qquad (6.21)$$

From this result we observe that as $t \to \infty$, the influence of the initial value decays exponentially and $\sigma_{V(t)}^2 \to \beta^2/2\alpha$. Thus, as $t \to \infty$, $V(t)$ converges exponentially to a Gaussian distribution with mean zero and variance $\beta^2/2\alpha$; that is,

$$\lim_{t \to \infty} V(t) \sim N\left(0,\ \beta^2/2\alpha\right) \qquad (6.22)$$

6.7.2 First Alternative Solution Method

Another method of solution of the OU SDE $dV(t) = -\alpha V(t)dt + \beta dB(t)$ is as follows. As in the previous method, we consider the function $f(t,\ V(t)) = e^{\alpha t}V(t)$ whose differential can be obtained directly as

$$d\left(e^{\alpha t}V(t)\right) = \alpha e^{\alpha t}V(t)dt + e^{\alpha t}dV(t)$$

$$= \alpha e^{\alpha t}V(t)dt + e^{\alpha t}\left\{-\alpha V(t)dt + \beta dB(t)\right\}$$

$$= e^{\alpha t}\beta dB(t)$$

which is the result we obtained earlier.

6.7.3 Second Alternative Solution Method

Because the OU process is a Gaussian process that is completely characterized by the mean and the variance, an alternative method of analyzing the process is that used in Gillespie (1996), which is obtained by rewriting the SDE for the process as follows:

$$dV(t) = V(t+dt) - V(t) = -\alpha V(t)dt + \beta dB(t)$$

Thus, taking expectations and remembering that $B(t)$ is a zero-mean process we obtain

$$E[V(t+dt)] - E[V(t)] = -\alpha E[V(t)]dt$$

$$\Rightarrow \lim_{dt \to 0} \frac{E[V(t+dt)] - E[V(t)]}{dt} = \frac{dE[V(t)]}{dt} - \alpha E[V(t)]$$

The solution to the second equation is

$$E[V(t)] = v_0 e^{-\alpha t} \quad t \geq 0$$

Also, because $V(t+dt) = V(t) - \alpha V(t)dt + \beta dB(t)$, taking the square of both sides gives

$$V^2(t+dt) = V^2(t) + \alpha^2 V^2(t)\{dt\}^2 + \beta^2 \{dB(t)\}^2 - 2\alpha V^2(t)dt + 2\beta V(t)dB(t)$$
$$- 2\alpha\beta V(t)dB(t)dt$$

Taking expectations of both sides and recalling that $E[\{dB(t)\}^2] = \sigma^2_{dB(t)} = dt$ and that $E[dB(t)] = 0$, we obtain

$$E[V^2(t+dt)] = E[V^2(t)] + \alpha^2 E[V^2(t)]\{dt\}^2 + \beta^2 dt - 2\alpha E[V^2(t)]dt$$

where we have assumed that $V(t)$ is statistically independent of $B(u)$ for all $t \leq u$ so that $E[V(t)dB(t)] = E[V(t)]E[dB(t)] = 0$. Thus, we obtain

$$E[V^2(t+dt)] - E[V^2(t)] = E[V^2(t)]\{\alpha^2 dt - 2\alpha\}dt + \beta^2 dt$$
$$\frac{E[V^2(t+dt)] - E[V^2(t)]}{dt} = E[V^2(t)]\{\alpha^2 dt - 2\alpha\} + \beta^2$$

Taking the limits as $dt \to 0$ we obtain

$$\lim_{dt \to 0} \frac{E[V^2(t+dt)] - E[V^2(t)]}{dt} = \frac{dE[V^2(t)]}{dt} = \beta^2 - 2\alpha E[V^2(t)]$$

If we assume that $V^2(0) = v_0^2$, then the solution to the second equation is given by

$$E\left[V^2(t)\right] = v_0^2 + \frac{\beta^2}{2\alpha}\left\{1 - e^{-2\alpha t}\right\}$$

Thus, using the previous result for $E[V(t)]$, the variance of $V(t)$ is given by

$$\text{Var}\left\{V(t)\right\} = E\left[V^2(t)\right] - \left\{E\left[V(t)\right]\right\}^2 = \frac{\beta^2}{2\alpha}\left\{1 - e^{-2\alpha t}\right\}$$

This means that

$$V(t) \sim N\left(v_0 e^{-\alpha t}, \frac{\beta^2}{2\alpha}\left\{1 - e^{-2\alpha t}\right\}\right)$$

as we obtained earlier.

6.8 MEAN-REVERTING ORNSTEIN–UHLENBECK PROCESS

A mean-reverting process is a process that, over time, tends to drift toward its long-term mean. The differential formula for a mean-reverting process $\{Y_t\}$ is as follows:

$$dY_t = \theta(K - Y_t)dt + \sigma dW_t$$

where $\theta > 0$ is the *rate of mean reversion* and K is the value around which Y_t tends to oscillate. The coefficient of dt is called the *drift term*. Specifically, when $Y_t > K$, the drift term is negative, which results in the process pulling back toward the equilibrium level. Similarly, when $Y_t < K$, the drift term is positive, which results in Y_t pulling up to the higher equilibrium level. Thus, another way to define a mean-reverting process is one whose changes in its values are negatively correlated. In this way, the process does not wander off to infinity. Instead it always tries to come back to a well-defined asymptotic mean value. For this reason, it is sometimes used to model processes such as prices, interest rates, and volatilities that tend to return to a mean or average value after reaching extremes.

A mean-reverting OU process $X(t)$ is the solution to the following SDE:

$$dX(t) = \alpha\left\{\mu - X(t)\right\}dt + \beta dB(t)$$

$$(6.23)$$

where $B(t)$ is the standard Brownian process, μ is the long-run mean of $X(t)$, and α is the rate of mean reversion.

As in previous sections, we solve this equation by considering $e^{\alpha t}X(t)$ and taking the differential

$$d\left(e^{\alpha t}X(t)\right) = \alpha e^{\alpha t}X(t)dt + e^{\alpha t}dX(t) = \alpha e^{\alpha t}X(t)dt + e^{\alpha t}\left[\alpha\{\mu - X(t)\}dt + \beta dB(t)\right]$$
$$= \alpha \mu e^{\alpha t}dt + \beta e^{\alpha t}dB(t)$$

Given that $X(0)=x_0$ we obtain the solution as

$$e^{\alpha t}X(t) = x_0 + \int_0^t \alpha \mu e^{\alpha u}du + \int_0^t \beta e^{\alpha u}dB(u)$$

which gives

$$X(t) = x_0 e^{-\alpha t} + \int_0^t \alpha \mu e^{-\alpha(t-u)}du + \int_0^t e^{-\alpha(t-u)}\beta dB(u)$$
$$= x_0 e^{-\alpha t} + \mu\left(1 - e^{-\alpha t}\right) + \int_0^t e^{-\alpha(t-u)}\beta dB(u)$$
$$= \mu + e^{-\alpha t}\left(x_0 - \mu\right) + \int_0^t e^{-\alpha(t-u)}\beta dB(u)$$

Because $B(t)$ is a Brownian motion, we have that the mean of $X(t)$ is

$$E[X(t)] = \mu + e^{-\alpha t}\left(x_0 - \mu\right) + E\left[\int_0^t e^{-\alpha(t-u)}\beta dB(u)\right]$$
$$= \mu + e^{-\alpha t}\left(x_0 - \mu\right)$$

Using the Ito isometry we obtain the variance of $X(t)$ as

$$\sigma_{X(t)}^2 = E\left[\left\{\int_0^t \beta e^{-\alpha(t-u)}dB(u)\right\}^2\right] = \int_0^t \left\{\beta e^{-\alpha(t-u)}\right\}^2 du = \beta^2 \int_0^t e^{-2\alpha(t-u)}du$$
$$= \frac{\beta^2}{2\alpha}\left(1 - e^{-2\alpha t}\right)$$

Thus, we have that

$$X(t) \sim N\left(\mu + e^{-\alpha t}\left(x_0 - \mu\right), \frac{\beta^2}{2\alpha}\left(1 - e^{-2\alpha t}\right)\right) \tag{6.24}$$

and

$$\lim_{t\to\infty} X(t) \sim N\left(\mu, \frac{\beta^2}{2\alpha}\right) \tag{6.25}$$

The difference between this process and the standard OU process is that as $t \to \infty$ the mean of the mean-reverting scheme is nonzero; in fact, it is μ. In the case of the standard OU process, the mean is 0.

6.9 SUMMARY

Stochastic calculus is used to model systems that have a random behavior. It is particularly applied to Brownian motion. In this chapter, we have presented the basic principles of stochastic calculus including the Ito integral and the Ito's formula, which we have used to solve SDEs for the geometric Brownian motion, the OU process and the mean-reverting OU.

CHAPTER 7

DIFFUSION PROCESSES

7.1 INTRODUCTION

Diffusion processes are used to model the movement of many objects in an environment or medium. The objects can be as small as basic particles in physics, bacteria, molecules, or cells; or very large objects like animals, plants, or certain types of events like epidemics and rumors. Thus, diffusion processes are important in many applications in science and engineering. Diffusion occurs when a system is not in equilibrium and random motion tends to bring everything toward uniformity. For example, diffusion enables heat to flow from a hot part of a conductor to a cold part of the conductor. Similarly, when we put a drop of dye in a jar of water, the dye spreads throughout the water, and we can say that the dye *diffused* through the water. Another example is when you are cooking food and the smell of the food spreads from the oven throughout the kitchen, and eventually throughout your home. In this case, the odor *diffused* through the air. It is diffusion that enables the heat from a furnace to warm a house; this is by convection.

In the preceding examples, the molecules diffused away from their original source because molecules always diffuse along their *concentration gradient*. This means that they diffuse from where they are in high concentration to where they are in low concentration. Once they have diffused to even out the concentration everywhere, the concentration gradient is now 0, and the molecules do not move in any particular

Elements of Random Walk and Diffusion Processes, First Edition. Oliver C. Ibe.
© 2013 John Wiley & Sons, Inc. Published 2013 by John Wiley & Sons, Inc.

direction anymore; they can now move randomly in all directions. Thus, we can define diffusion as the spontaneous "spreading" of particles from a region of high concentration to one of lower concentration.

Diffusion processes are used to model the price movements of financial instruments. The Black–Scholes model for pricing options assumes that the underlying instrument follows a traditional diffusion process with small, continuous, random movements.

Diffusion processes are usually described via partial differential equations. This makes it attractive for many discrete processes to be approximated by a diffusion process because partial differential equations are generally easier to solve than the differential-difference equations that are often used to describe the evolution of these processes.

7.2 MATHEMATICAL PRELIMINARIES

Consider a continuous-time continuous-state Markov process $\{X(t),\ t \geq 0\}$ whose transition probability distribution is given by

$$F(y, t; x, s) = P[X(t) \leq y \mid X(s) = x], \quad s < t \qquad (7.1)$$

If the derivative

$$f(y, t; x, s) = \frac{\partial}{\partial y} F(y, t; x, s) \qquad (7.2)$$

exists, then it is called the *transition density function* of the diffusion process. Since $\{X(t)\}$ is a Markov process, $f(y, t; x, s)$ satisfies the Chapman–Kolmogorov equation:

$$f(y, t; x, s) = \int_{-\infty}^{\infty} f(y, t; z, u) f(z, u; x, s) dz \qquad (7.3)$$

We assume that the process $\{X(t),\ t \geq 0\}$ satisfies the following conditions:

1. $P[|X(t+\Delta t) - X(t)| > \varepsilon | X(t)] = o(\Delta t)$, for $\varepsilon > 0$, which states that the sample path is continuous; alternatively we say that the process is continuous in probability.
2. $E\left[X(t+\Delta t) - X(t) \mid X(t) = x\right] = a(x, t)\Delta t + o(\Delta t)$ so that

$$\lim_{\Delta t \to 0} \frac{E\left[X(t+\Delta t) - X(t) \mid X(t) = x\right]}{\Delta t}$$

$$= \lim_{\Delta t \to 0} \frac{1}{\Delta t} \int_{-\infty}^{\infty} (y - x) f(y, t+\Delta t) \mid x, t) \, dy = a(x, t)$$

3. $E\left[\{X(t+\Delta t) - X(t)\}^2 \mid X(t) = x\right] = b(x, t)\Delta t + o(\Delta t)$ is finite so that

$$\lim_{\Delta t \to 0} \frac{E[\{X(t+\Delta t)-X(t)\}^2 \,|\, X(t)=x]}{\Delta t}$$

$$= \lim_{\Delta t \to 0} \frac{1}{\Delta t} \int_{-\infty}^{\infty} (y-x)^2 f(y, t+\Delta t)|x, t) dy = b(x, t)$$

A Markov process that satisfies these three conditions is called a diffusion process. The function $a(x, t)$ is called the *instantaneous* (or infinitesimal) *mean* (or drift) of $X(t)$, and the function $b(x, t)$ is called the *instantaneous* (or infinitesimal) *variance* of $X(t)$. Let the small increment in $X(t)$ over any small interval dt be denoted by $dX(t)$. Then it can be shown that if $B(t)$ is a standard Brownian motion we can incorporate the aforementioned properties into the following stochastic differential equation:

$$dX(t) = a(x, t)dt + [b(x, t)]^{1/2} dB(t) \qquad (7.4)$$

where $dB(t)$ is the increment of $B(t)$ over the small interval $(t, t+\Delta t)$.

7.3   DIFFUSION ON ONE-DIMENSIONAL RANDOM WALK

The diffusion equation is a partial differential equation that describes the density fluctuations in a material undergoing diffusion. Diffusion can be obtained as a limit of the random walk. Consider a one-dimensional random walk where in each interval of length Δt the process makes a movement of length Δx with probability p or a movement of length $-\Delta x$ with probability $q = 1-p$. Let $P[x, t; x_0, t_0]$ denote the probability that the process is at x at time t, given that it was at x_0 at time t_0 where $\Delta x = x - x_0$ and $\Delta t = t - t_0$. Assume that p and q are independent of x and t. Then we have that

$$P[x, t; x_0, t_0] = pP[x - \Delta x, t - \Delta t; x_0, t_0] + qP[x + \Delta x, t - \Delta t; x_0, t_0]$$

Since $P[x, t; x_0, t_0] = f(x, t; x_0, t_0)\Delta x$, we have that

$$f(x, t; x_0, t_0)\Delta x = pf(x - \Delta x, t - \Delta t; x_0, t_0)\Delta x + qf(x + \Delta x, t - \Delta t; x_0, t_0)\Delta x$$

Thus, we obtain

$$f(x, t; x_0, t_0) = pf(x - \Delta x, t - \Delta t; x_0, t_0) + qf(x + \Delta x, t - \Delta t; x_0, t_0)$$

We assume that $f(x, t; x_0, t_0)$ is a smooth function of its arguments so that we can expand it in a Taylor series in x and t. Thus, we have that

$$f(x,t;x_0,t_0) = p\left\{ f(x,t;x_0,t_0) - \Delta x \frac{\partial f}{\partial x} - \Delta t \frac{\partial f}{\partial t} + \frac{(\Delta x)^2}{2} \frac{\partial^2 f}{\partial x^2} + \frac{(\Delta t)^2}{2} \frac{\partial^2 f}{\partial t^2} \right. $$

$$\left. + (\Delta x)(\Delta t) \frac{\partial^2 f}{\partial x \partial t} \right\}$$

$$+ q\left\{ f(x,t;x_0,t_0) + \Delta x \frac{\partial f}{\partial x} - \Delta t \frac{\partial f}{\partial t} + \frac{(\Delta x)^2}{2} \frac{\partial^2 f}{\partial x^2} + \frac{(\Delta t)^2}{2} \frac{\partial^2 f}{\partial t^2} \right.$$

$$\left. - (\Delta x)(\Delta t) \frac{\partial^2 f}{\partial x \partial t} \right\}$$

$$+ o\left(\{\Delta x\}^3 \right) + o\left(\{\Delta t\}^3 \right)$$

$$= f(x,t;x_0,t_0) - \Delta t \frac{\partial f}{\partial t} - (p-q)\Delta x \frac{\partial f}{\partial x} + \frac{(\Delta x)^2}{2} \frac{\partial^2 f}{\partial x^2} + \frac{(\Delta t)^2}{2} \frac{\partial^2 f}{\partial t^2}$$

$$+ (p-q)(\Delta x)(\Delta t) \frac{\partial^2 f}{\partial x \partial t} + o\left(\{\Delta x\}^3 \right) + o\left(\{\Delta t\}^3 \right)$$

From this we obtain

$$\frac{\partial f}{\partial t} = -\frac{(p-q)\Delta x}{\Delta t} \frac{\partial f}{\partial x} + \frac{(\Delta x)^2}{2\Delta t} \frac{\partial^2 f}{\partial x^2} + \frac{\Delta t}{2} \frac{\partial^2 f}{\partial t^2}$$

$$+ (p-q)\Delta x \frac{\partial^2 f}{\partial x \partial t} + o\left(\frac{\{\Delta x\}^3}{\Delta t} \right) + o\left(\{\Delta t\}^2 \right)$$

Assume that

$$\mu = \lim_{\Delta t, \Delta x \to 0} \frac{(p-q)\Delta x}{\Delta t}$$

$$D_0 = \lim_{\Delta t, \Delta x \to 0} \left\{ \frac{(\Delta x)^2}{2\Delta t} \right\} \tag{7.5}$$

Then in the limit as Δx, $\Delta t \to 0$ we have that

$$\frac{\partial f}{\partial t} = -\mu \frac{\partial f}{\partial x} + D_0 \frac{\partial^2 f}{\partial x^2} \tag{7.6}$$

This is called the *forward diffusion equation* because it involves differentiation in x. This equation is commonly called the *Fokker–Planck equation*. The constant D_0 is called the *diffusion constant* or *diffusivity*; it describes how fast or slow an object diffuses. The equation obtained when $\mu = 0$ (i.e., when $p = q = 1/2$) is known as the *heat equation*, which is

$$\frac{\partial f}{\partial t} = D_0 \frac{\partial^2 f}{\partial x^2} \tag{7.7}$$

Similarly, when $D_0 = 0$, we obtain the wave equation:

$$\frac{\partial f}{\partial t} + \mu \frac{\partial f}{\partial x} = 0 \tag{7.8}$$

The *backward diffusion equation*, which involves differentiation in x_0, can be obtained as follows:

$$\frac{\partial f}{\partial t} = \mu \frac{\partial f}{\partial x_0} + D_0 \frac{\partial^2 f}{\partial x_0^2} \tag{7.9}$$

The difference between the forward diffusion equation and the backward diffusion equation is as follows. If at time t the state of the system is x, the forward equation wants to know about the distribution of the state at a future time $s > t$ (hence, the term "forward"). On the other hand, the backward equation is useful when we want to address the following question: Given that the system at a future time s has a particular behavior, what can we say about the distribution at time $t < s$? In this case, the partial differential equation is integrated backward in time, from s to t (hence, the term "backward"). Thus, if the forward diffusion equation models the evolution, say, of the temperature from its initial values, the backward equation addresses the issue of determining the initial distribution from which the current temperature could have evolved.

The term $\mu(\partial f/\partial x)$ is called a *convection term*, while the term $D_0(\partial^2 f/\partial x^2)$ is called the diffusion term. Thus, Equation 7.9 is often called a *diffusion equation with convection*, or a *convection–diffusion equation*.

A physically motivated derivation for the diffusion equation starts with the Fick's laws. As discussed earlier, diffusion is essentially the movement of a substance from a region of high concentration to a region of low concentration. Given enough time, this movement will eventually result in homogeneity within the entire region causing the movement due to diffusion to stop. The first Fick's law states that the magnitude of the diffusive flux of the substance across a given plane from the region of high concentration to the region of low concentration is proportional to the concentration gradient across the plane. That is, if ϕ is the concentration of the diffusive substance and J is the diffusion flux (or the amount of substance per unit area per unit time), then

$$J = -D_0 \frac{\partial \phi}{\partial x} \tag{7.10}$$

where D_0 is the diffusion constant discussed earlier. The negative sign of the right-hand side of the equation indicates that the substance is flowing in the direction of lower concentration.

Note that this law does not consider the fact that the gradient and local concentration of the substance decrease as time increases. The flux of substance entering a section of a bar with a concentration gradient is different from the flux leaving the same section. From the law of conservation of matter, the difference between the two fluxes must lead to a change in the concentration of substance within the section. This is precisely *Fick's second law*, which states that the change in substance concentration over time is equal to the change in local diffusion flux. That is,

$$\frac{\partial \phi}{\partial t} = -\frac{\partial J}{\partial x}$$

Combining this with Fisk's first law, we obtain

$$\frac{\partial \phi}{\partial t} = -\frac{\partial J}{\partial x} = \frac{\partial}{\partial x}\left(D_0 \frac{\partial \phi}{\partial x}\right) = D_0 \frac{\partial^2 \phi}{\partial x^2} \tag{7.11}$$

7.3.1 Alternative Derivation

We start with the random walk expression

$$P[x, t] = pP[x - \Delta x, t - \Delta t] + qP[x + \Delta x, t - \Delta t]$$

Since $p + q = 1$, we have that

$$P[x, t] - P[x, t - \Delta t] = pP[x - \Delta x, t - \Delta t] + qP[x + \Delta x, t - \Delta t] - (p + q)P[x, t - \Delta t]$$

From the Taylor's series expansion we have that

$$P[x, t] - \left\{P[x, t] - \Delta t \frac{\partial P}{\partial t}\right\} = q\left\{P[x, t] + \Delta x \frac{\partial P}{\partial x} - \Delta t \frac{\partial P}{\partial t} + \frac{1}{2}(\Delta x)^2 \frac{\partial^2 P}{\partial x^2} + \frac{1}{2}(\Delta t)^2 \frac{\partial^2 P}{\partial t^2}\right\}$$

$$-q\left\{P[x, t] - \Delta t \frac{\partial P}{\partial t} + \frac{1}{2}(\Delta t)^2 \frac{\partial^2 P}{\partial t^2}\right\}$$

$$-p\left\{P[x, t] - \Delta t \frac{\partial P}{\partial t} + \frac{1}{2}(\Delta t)^2 \frac{\partial^2 P}{\partial t^2}\right\}$$

$$+p\left\{P[x, t] - \Delta x \frac{\partial P}{\partial x} - \Delta t \frac{\partial P}{\partial t} + \frac{1}{2}(\Delta x)^2 \frac{\partial^2 P}{\partial x^2} + \frac{1}{2}(\Delta t)^2 \frac{\partial^2 P}{\partial t^2}\right\}$$

$$+o\left(\{\Delta x\}^3\right) + o\left(\{\Delta t\}^3\right)$$

Thus, we have that

$$
\begin{aligned}
\frac{\partial P}{\partial t} &= q\left\{\frac{\Delta x}{\Delta t}\frac{\partial P}{\partial x} + \frac{(\Delta x)^2}{2\Delta t}\frac{\partial^2 P}{\partial x^2}\right\} - p\left\{\frac{\Delta x}{\Delta t}\frac{\partial P}{\partial x} - \frac{(\Delta x)^2}{2\Delta t}\frac{\partial^2 P}{\partial x^2}\right\} \\
&= \frac{(\Delta x)^2}{2\Delta t}\frac{\partial^2 P}{\partial x^2} - \frac{(p-q)\Delta x}{\Delta t}\frac{\partial P}{\partial x}
\end{aligned}
$$

As discussed earlier, if we define

$$
\mu = \lim_{\Delta t,\, \Delta x \to 0} \frac{(p-q)\Delta x}{\Delta t}
$$

$$
D_0 = \lim_{\Delta t,\, \Delta x \to 0} \left\{\frac{(\Delta x)^2}{2\Delta t}\right\}
$$

then we obtain

$$
\frac{\partial P}{\partial t} = D_0 \frac{\partial^2 P}{\partial x^2} - \mu \frac{\partial P}{\partial x}
$$

which is the same equation as Equation 7.6.

7.4   EXAMPLES OF DIFFUSION PROCESSES

The following are some examples of the diffusion process. These processes differ only in their values of the infinitesimal mean and infinitesimal variance.

7.4.1   Brownian Motion

The Brownian motion is a diffusion process on the interval $(-\infty, \infty)$ with zero mean and constant variance. That is, for the standard Brownian motion, $\mu=0$ and $D_0=\sigma^2/2$, where $\sigma^2>0$ is the variance. Thus, the forward diffusion equation becomes

$$
\frac{\partial f}{\partial t} = \frac{\sigma^2}{2}\frac{\partial^2 f}{\partial x^2} \tag{7.12}
$$

As stated earlier, this equation is an example of a heat equation, which describes the variation in temperature as heat diffuses through an isotropic and homogeneous physical medium in one-dimensional space. Similarly, the backward diffusion equation becomes

$$
\frac{\partial f}{\partial t} = -\frac{\sigma^2}{2}\frac{\partial^2 f}{\partial x_0^2} \tag{7.13}
$$

We can solve the forward diffusion equation as follows. If we define $\lambda^2 = \sigma^2/2$, the forward diffusion equation becomes

$$\frac{\partial}{\partial t} f(x, t) = \lambda^2 \frac{\partial^2}{\partial x^2} f(x, t) \quad 0 \le x \le L; 0 \le t \le T$$

where

$$f(x, t) = \frac{\partial}{\partial x} P[X(t) \le x]$$

Assume that the initial condition is $f(x, 0) = \phi(x)$, $0 \le x \le L$, and the boundary conditions are $f(0, t) = f(L, t) = 0$, $0 \le t \le T$. If we assume that $f(x, t)$ is a separable function of t and x, then we can write

$$f(x, t) = g(t)h(x)$$

with the boundary conditions $h(0) = h(L) = 0$. Thus, the differential equation becomes

$$\frac{dg(t)}{dt} h(x) = \lambda^2 g(t) \frac{d^2 h(x)}{dx^2}$$

which gives

$$\frac{1}{\lambda^2} \frac{dg(t)}{dt} \frac{1}{g(t)} = \frac{d^2 h(x)}{dx^2} \frac{1}{h(x)}$$

Because the left side of the equation is a function of t alone and the right side is a function of x alone, the equality can be satisfied only if the two sides are equal to some constant m, say; that is,

$$\frac{dg(t)}{dt} = m\lambda^2 g(t) \quad 0 \le t \le T$$

$$\frac{d^2 h(x)}{dx^2} = mh(x) \quad 0 \le x \le L$$

Using the method of characteristic equation, the solutions to these equations are given by

$$g(t) = C_0 e^{m\lambda^2 t}$$

$$h(x) = C_1 e^{x\sqrt{m}} + C_2 e^{-x\sqrt{m}}$$

where C_0, C_1, and C_2 are constants. To avoid a trivial solution obtained when $m=0$, and to obtain a solution that does not increase exponentially with t, we require that $m<0$. Thus, if we define $m=-p^2$, the solutions become

$$g(t) = C_0 e^{-\lambda^2 p^2 t}$$

$$h(x) = C_1 e^{jpx} - C_2 e^{-jpx} = B_1 \sin(px) + B_2 \cos(px)$$

$$f(x, t) = C_0 e^{-\lambda^2 p^2 t}\{B_1 \sin(px) + B_2 \cos(px)\}$$

From the boundary condition $f(0, t)=0$, we obtain $B_2=0$. Similarly, from the boundary condition $f(L, t)=0$, we obtain $B_1 \sin(pL)=0$, which gives $pL=k\pi$, $k=1, 2, \dots$. Thus,

$$p = \frac{k\pi}{L}, \quad k = 1, 2, \dots$$

We can then define the following functions:

$$g_k(t) = C_{0k} \exp\left\{-\frac{\lambda^2 k^2 \pi^2 t}{L^2}\right\}$$

$$h_k(x) = B_{1k} \sin\left\{\frac{k\pi x}{L}\right\}$$

$$f_k(x, t) = C_k \exp\left\{-\frac{\lambda^2 k^2 \pi^2 t}{L^2}\right\} \sin\left\{\frac{k\pi x}{L}\right\}$$

$$f(x, t) = \sum_{k=1}^{\infty} f_k(x, t) = \sum_{k=1}^{\infty} C_k \exp\left\{-\frac{\lambda^2 k^2 \pi^2 t}{L^2}\right\} \sin\left\{\frac{k\pi x}{L}\right\}$$

where $C_k = C_{0k} \times B_{1k}$. From the initial condition, we obtain

$$f(x, 0) = \phi(x) = \sum_{k=1}^{\infty} C_k \sin\left(\frac{k\pi x}{L}\right)$$

Because

$$\int_0^L \sin\left(\frac{m\pi u}{L}\right) \sin\left(\frac{k\pi u}{L}\right) du = \begin{cases} 0 & \text{if } k \neq m \\ \dfrac{L}{2} & \text{if } k = m \end{cases}$$

that is, the functions $\{\sin(k\pi u)/L\}$ are orthogonal on the interval $0 \le u \le L$, we obtain

$$C_k = \frac{2}{L}\int_0^L \phi(u)\sin\left(\frac{k\pi u}{L}\right)du$$

This means that

$$f(x,t) = \frac{2}{L}\sum_{k=1}^{\infty}\left\{\int_0^L \phi(u)\sin\left(\frac{k\pi u}{L}\right)du\right\}\exp\left\{-\frac{\lambda^2 k^2 \pi^2 t}{L^2}\right\}\sin\left\{\frac{k\pi x}{L}\right\} \quad (7.14)$$

7.4.2 Brownian Motion with Drift

In this process, $a(x,t)=\mu$ and $b(x,t)=\sigma^2$, where μ is the drift rate. The forward diffusion equation becomes

$$\frac{\partial f}{\partial t} = -\mu\frac{\partial f}{\partial x} + \frac{\sigma^2}{2}\frac{\partial^2 f}{\partial x^2} \quad (7.15)$$

Similarly, the backward diffusion equation becomes

$$\frac{\partial f}{\partial t} = -\mu\frac{\partial f}{\partial x_0} - \frac{\sigma^2}{2}\frac{\partial^2 f}{\partial x_0^2} \quad (7.16)$$

7.5 CORRELATED RANDOM WALK AND THE TELEGRAPH EQUATION

Let $P(x,t)$ denote the probability that a walker performing a one-dimensional correlated random walk reaches x from 0 at a time t. Let $R(x,t)$ denote the probability that the walker arrives at x from the right, and let $L(x,t)$ denote the probability that the walker reaches x from the left. Then we have that

$$P(x,t) = R(x,t) + L(x,t) \quad (7.17)$$

We analyze the problem using a method used in Jain (1971). Assume that the walker takes steps of length Δx and the time between two consecutive steps is Δt. Now,

$$
\begin{aligned}
R(x,t+\Delta t) &= p_1 R(x-\Delta x,t) + q_2 L(x-\Delta x,t)\\
&= p_1\{P(x-\Delta x,t) - L(x-\Delta x,t)\} + q_2 L(x-\Delta x,t)\\
&= p_1 P(x-\Delta x,t) - (p_1 - q_2)L(x-\Delta x,t)\\
L(x,t+\Delta t) &= p_2 L(x+\Delta x,t) + q_1 R(x+\Delta x,t)\\
&= p_2\{P(x+\Delta x,t) - R(x+\Delta x,t)\} + q_1 R(x+\Delta x,t)\\
&= p_2 P(x+\Delta x,t) - (p_1 - q_2)R(x+\Delta x,t)
\end{aligned}
\quad (7.18)
$$

where the last equality follows from the fact that $p_2 - q_1 = (1 - q_2) - (1 - p_1) = p_1 - q_2$. Also,

$$
\begin{aligned}
R(x + \Delta x, t) + L(x - \Delta x, t) &= \{ p_1 R(x, t - \Delta t) + q_2 L(x, t - \Delta t) \} \\
&\quad + \{ p_2 L(x, t - \Delta t) + q_1 R(x, t - \Delta t) \} \\
&= R(x, t - \Delta t) + L(x, t - \Delta t) \\
&= P(x, t - \Delta t)
\end{aligned} \tag{7.19}
$$

From Equations 7.17, 7.18 and 7.19 we have that

$$
\begin{aligned}
P(x, t + \Delta t) &= p_1 P(x - \Delta x, t) + p_2 P(x + \Delta x, t) \\
&\quad - (p_1 - q_2)\{ R(x + \Delta x, t) + L(x - \Delta x, t) \} \\
&= p_1 P(x - \Delta x, t) + p_2 P(x + \Delta x, t) \\
&\quad - (p_1 - q_2) P(x, t - \Delta t)
\end{aligned} \tag{7.20}
$$

Using the Taylor series expansion we have that

$$
P(x, t + \Delta t) = P(x, t) + \Delta t \frac{\partial P}{\partial t} + \frac{1}{2}(\Delta t)^2 \frac{\partial^2 P}{\partial t^2} + o\left(\{\Delta t\}^3\right)
$$

$$
P(x, t - \Delta t) = P(x, t) - \Delta t \frac{\partial P}{\partial t} + \frac{1}{2}(\Delta t)^2 \frac{\partial^2 P}{\partial t^2} + o\left(\{\Delta t\}^3\right)
$$

$$
P(x + \Delta x, t) = P(x, t) + \Delta x \frac{\partial P}{\partial x} + \frac{1}{2}(\Delta x)^2 \frac{\partial^2 P}{\partial x^2} + o\left(\{\Delta x\}^3\right)
$$

$$
P(x - \Delta x, t) = P(x, t) - \Delta x \frac{\partial P}{\partial x} + \frac{1}{2}(\Delta x)^2 \frac{\partial^2 P}{\partial x^2} + o\left(\{\Delta x\}^3\right)
$$

Thus, substituting in Equation 7.20, we obtain

$$
\begin{aligned}
P(x, t) + \Delta t \frac{\partial P}{\partial t} + \frac{1}{2}(\Delta t)^2 \frac{\partial^2 P}{\partial t^2} &= p_1 \left\{ P(x, t) - \Delta x \frac{\partial P}{\partial x} + \frac{1}{2}(\Delta x)^2 \frac{\partial^2 P}{\partial x^2} \right\} \\
&\quad + p_2 \left\{ P(x, t) + \Delta x \frac{\partial P}{\partial x} + \frac{1}{2}(\Delta x)^2 \frac{\partial^2 P}{\partial x^2} \right\} \\
&\quad - (p_1 - q_2) \left\{ P(x, t) - \Delta t \frac{\partial P}{\partial t} + \frac{1}{2}(\Delta t)^2 \frac{\partial^2 P}{\partial t^2} \right\} \\
&\quad + o\left(\{\Delta t\}^3\right) + o\left(\{\Delta x\}^3\right)
\end{aligned}
$$

This gives

$$\Delta t \frac{\partial P}{\partial t}(1 - p_1 + q_2) + \frac{1}{2}(\Delta t)^2 \frac{\partial^2 P}{\partial t^2}(1 + p_1 - q_2)$$

$$= (p_1 + p_2)\frac{1}{2}(\Delta x)^2 \frac{\partial^2 P}{\partial x^2} - (p_1 - p_2)\Delta x \frac{\partial P}{\partial x} + o\left(\{\Delta x\}^3\right) + o\left(\{\Delta t\}^3\right)$$

That is,

$$\frac{\partial^2 P}{\partial t^2} + \frac{2(q_1 + q_2)}{(p_1 + p_2)\Delta t}\frac{\partial P}{\partial t} = \frac{(p_1 + q_2)(\Delta x)^2}{(1 + p_1 - q_2)(\Delta t)^2}\frac{\partial^2 P}{\partial x^2}$$

$$- \frac{2(p_1 - q_2)\Delta x}{(1 + p_1 - q_2)(\Delta t)^2}\frac{\partial P}{\partial x} + o\left(\{\Delta x\}^3\right) + o\left(\{\Delta t\}^3\right)$$

$$= \frac{(p_1 + q_2)}{(1 + p_1 - q_2)}\left(\frac{\Delta x}{\Delta t}\right)^2 \frac{\partial^2 P}{\partial x^2}$$

$$- \frac{2(p_1 - p_2)}{(1 + p_1 - q_2)\Delta x}\left(\frac{\Delta x}{\Delta t}\right)^2 \frac{\partial P}{\partial x}$$

$$+ o(\{\Delta t\}) + o\left(\frac{\{\Delta x\}^3}{\{\Delta t\}^2}\right)$$

$$= \left(\frac{\Delta x}{\Delta t}\right)^2 \frac{\partial^2 P}{\partial x^2} - \frac{2(p_1 - p_2)}{(p_1 + p_2)\Delta x}\left(\frac{\Delta x}{\Delta t}\right)^2 \frac{\partial P}{\partial x}$$

$$+ o(\{\Delta t\}) + o\left(\frac{\{\Delta x\}^3}{\{\Delta t\}^2}\right)$$

Let $v = \lim\limits_{\Delta x, \Delta t \to 0} \{\Delta x / \Delta t\}$ be the velocity of the walker and let the constants T and D_1 be defined as follows:

$$\frac{1}{T} = \lim_{\Delta t \to 0}\frac{(q_1 + q_2)}{\Delta t(p_1 + p_2)}$$

$$D_1 = \lim_{\Delta x \to 0}\frac{p_1 - p_2}{(p_2 + p_2)\Delta x} \tag{7.21}$$

Then, in the limit $\Delta t, \Delta x \to 0$, we obtain

$$\frac{\partial^2 P}{\partial t^2} + \frac{2}{T}\frac{\partial P}{\partial t} = v^2 \left\{\frac{\partial^2 P}{\partial x^2} - 2D_1 \frac{\partial P}{\partial x}\right\} \tag{7.22}$$

which is the telegraph equation. When $p_1 = p_2$, we have that $D_1 = 0$, and we obtain the telegraph equation without leakage, namely

$$\frac{\partial^2 P}{\partial t^2} + \frac{2}{T}\frac{\partial P}{\partial t} = v^2 \frac{\partial^2 P}{\partial x^2} \tag{7.23}$$

which is the equation of wave motion with speed v. Thus, the telegraph equation features the characteristics of both the diffusion process and wave motion. The physical difference between the telegraph equation and the diffusion equation is that the velocity of dispersion is finite in the telegraph equation but infinite in the diffusion equation.

7.6 DIFFUSION AT FINITE SPEED

Diffusion is the result of particles making a random walk that involves a sequence of small random steps. While each step of the random walk is made at a finite speed, the resulting diffusion occurs at an infinite speed. This paradox can be resolved by modeling the limiting walk by the telegraph equation, as discussed earlier. We adopt the method used in Keller (2004) to give an alternative method of deriving Equation 7.23.

Consider a random walker who takes steps of length Δx with probability p in the positive direction and steps of length $-\Delta x$ with probability $q = 1 - p$ (obviously in the negative direction). The interval between two consecutive steps is Δt, and the steps are taken independently of each other. Let $P(x, t)$ denote the probability that the walker is at location x at time t, where the location is measured from the origin where he started from. Then we have that

$$P(x, t + \Delta t) = pP(x - \Delta x, t) + qP(x + \Delta x, t)$$

We assume that $P(x, t)$ is a smooth function of its arguments so we can use the Taylor series expansion in x and t and omit the derivatives of higher order than 2 as follows:

$$P(x, t) + \Delta t \frac{\partial P}{\partial t} + \frac{(\Delta t)^2}{2}\frac{\partial^2 P}{\partial t^2} = p\left\{ P(x, t) - \Delta x \frac{\partial P}{\partial x} + \frac{(\Delta x)^2}{2}\frac{\partial^2 P}{\partial x^2}\right\}$$
$$+ q\left\{ P(x, t) + \Delta x \frac{\partial P}{\partial x} + \frac{(\Delta x)^2}{2}\frac{\partial^2 P}{\partial x^2}\right\}$$
$$+ o\left((\Delta t)^3\right) + o\left((\Delta x)^3\right)$$

This gives

$$\frac{(\Delta t)^2}{2}\frac{\partial^2 P}{\partial t^2} + \Delta t \frac{\partial P}{\partial t} = \frac{(\Delta x)^2}{2}\frac{\partial^2 P}{\partial x^2} - (p - q)\Delta x \frac{\partial P}{\partial x} + o\left((\Delta t)^3\right) + o\left((\Delta x)^3\right)$$

Thus,

$$\frac{\partial^2 P}{\partial t^2} + \frac{2}{\Delta t}\frac{\partial P}{\partial t} = \left(\frac{\Delta x}{\Delta t}\right)^2\frac{\partial^2 P}{\partial x^2} - \frac{2(p-q)}{\Delta x}\left(\frac{\Delta x}{\Delta t}\right)^2\frac{\partial P}{\partial x} + o(\{\Delta t\}) + o\left(\frac{\{\Delta x\}^3}{\{\Delta t\}^2}\right)$$

If we define $v = \lim_{\Delta x,\, \Delta t \to 0}\{\Delta x / \Delta t\}$ as the velocity of the walker,

$$D_2 = \lim_{\Delta x \to 0}\frac{(p-q)}{\Delta x}$$

$$\frac{1}{T} = \lim_{\Delta t \to 0}\frac{1}{\Delta t} \tag{7.24}$$

then we have that in the limit as $\Delta t,\, \Delta x \to 0$,

$$\frac{\partial^2 P}{\partial t^2} + \frac{2}{T}\frac{\partial P}{\partial t} = v^2\left\{\frac{\partial^2 P}{\partial x^2} - 2D_2\frac{\partial P}{\partial x}\right\} \tag{7.25}$$

which we obtained earlier as the telegraph equation. For a symmetric random walk where $p=q=1/2$, which means that $D_2=0$, we obtain

$$\frac{\partial^2 P}{\partial t^2} + \frac{2}{T}\frac{\partial P}{\partial t} = v^2\frac{\partial^2 P}{\partial x^2} \tag{7.26}$$

7.7 DIFFUSION ON SYMMETRIC TWO-DIMENSIONAL LATTICE RANDOM WALK

As discussed in Chapter 4, in the symmetric two-dimensional random walk, the walker takes one step either in the positive x direction, negative x direction, positive y direction, or negative y direction with equal probability. The step size is fixed, and each step is independent of the other steps.

Let $P(x, y, t)$ denote the probability that the walker is at location (x, y) at time t. We assume that at each time interval Δt the walker takes a step of length h in the north, south, east, or west with equal probability $1/4$. Thus, we obtain the following equation:

$$P(x, y, t) = \frac{1}{4}\{P(x-h, y, t-\Delta t) + P(x+h, y, t-\Delta t) + P(x, y-h, t-\Delta t)$$
$$+ P(x, y+h, t-\Delta t)\}$$

From Taylor's series we have that

$$P(x+h, y, t - \Delta t) = P(x, y, t) + \left(h \frac{\partial P}{\partial x} - \Delta t \frac{\partial P}{\partial t} \right)$$
$$+ \frac{1}{2} \left(h^2 \frac{\partial^2 P}{\partial x^2} + (\Delta t)^2 \frac{\partial^2 P}{\partial t^2} - 2h\Delta t \frac{\partial^2 P}{\partial x \partial t} \right) + o(h^3) + o\left(\{\Delta t\}^3 \right)$$

$$P(x-h, y, t - \Delta t) = P(x, y, t) - \left(h \frac{\partial P}{\partial x} + \Delta t \frac{\partial P}{\partial t} \right)$$
$$+ \frac{1}{2} \left(h^2 \frac{\partial^2 P}{\partial x^2} + (\Delta t)^2 \frac{\partial^2 P}{\partial t^2} + 2h\Delta t \frac{\partial^2 P}{\partial x \partial t} \right) + o(h^3) + o\left(\{\Delta t\}^3 \right)$$

$$P(x, y+h, t - \Delta t) = P(x, y, t) + \left(h \frac{\partial P}{\partial y} - \Delta t \frac{\partial P}{\partial t} \right)$$
$$+ \frac{1}{2} \left(h^2 \frac{\partial^2 P}{\partial y^2} + (\Delta t)^2 \frac{\partial^2 P}{\partial t^2} - 2h\Delta t \frac{\partial^2 P}{\partial y \partial t} \right) + o(h^3) + o\left(\{\Delta t\}^3 \right)$$

$$P(x, y-h, t - \Delta t) = P(x, y, t) - \left(h \frac{\partial P}{\partial y} + \Delta t \frac{\partial P}{\partial t} \right)$$
$$+ \frac{1}{2} \left(h^2 \frac{\partial^2 P}{\partial y^2} + (\Delta t)^2 \frac{\partial^2 P}{\partial t^2} + 2h\Delta t \frac{\partial^2 P}{\partial y \partial t} \right) + o(h^3) + o\left(\{\Delta t\}^3 \right)$$

Thus, we obtain

$$\Delta t \frac{\partial P}{\partial t} = \frac{1}{4} \left(h^2 \frac{\partial^2 P}{\partial x^2} + h^2 \frac{\partial^2 P}{\partial y^2} + 2(\Delta t)^2 \frac{\partial^2 P}{\partial t^2} \right) + o(h^3) + o\left(\{\Delta t\}^3 \right)$$

that is,

$$\frac{\partial P}{\partial t} = \frac{h^2}{4\Delta t} \left(\frac{\partial^2 P}{\partial x^2} + \frac{\partial^2 P}{\partial y^2} \right) + 2\Delta t \frac{\partial^2 P}{\partial t^2} + \frac{o(h^3)}{\Delta t} + o\left(\{\Delta t\}^2 \right)$$

If we define

$$D_3 = \lim_{h, \Delta t \to 0} \left\{ \frac{h^2}{4\Delta t} \right\} \qquad (7.27)$$

as the diffusion constant, then in the limit as $h, \Delta t \to 0$, we obtain

$$\frac{\partial P}{\partial t} = D_3 \left(\frac{\partial^2 P}{\partial x^2} + \frac{\partial^2 P}{\partial y^2} \right) \qquad (7.28)$$

7.8   DIFFUSION APPROXIMATION OF THE PEARSON RANDOM WALK

Let $P(x, y, t)$ denote the probability that the walker is at location (x, y) at time t. Let the probability density function of the turn angle Θ be $f_\Theta(\theta) = 1/2\pi$, $0 \le \theta \le 2\pi$. Let the length of each step be r. Then at time $t - \Delta t$, the walker was at location $(x - r \sin\theta, y - r \cos\theta)$. Thus, the dynamics of the process becomes

$$P(x, y, t) = \int_0^{2\pi} P(x - r\sin\theta, y - r\cos\theta, t - \Delta t) f_\Theta(\theta) d\theta$$

Using the Taylor series expansion of the term within the integral we obtain

$$P = \int_0^{2\pi} \left\{ P - r\sin\theta \frac{\partial P}{\partial x} - r\cos\theta \frac{\partial P}{\partial y} - \Delta t \frac{\partial P}{\partial t} \right\} f_\Theta(\theta) d\theta$$

$$+ \int_0^{2\pi} \left\{ \frac{r^2 \sin^2\theta}{2} \frac{\partial^2 P}{\partial x^2} + \frac{r^2 \cos^2\theta}{2} \frac{\partial^2 P}{\partial y^2} + \frac{(\Delta t)^2}{2} \frac{\partial^2 P}{\partial t^2} \right\} f_\Theta(\theta) d\theta$$

$$+ \int_0^{2\pi} \left\{ r^2 \sin\theta \cos\theta \frac{\partial^2 P}{\partial x \partial y} + r\sin\theta \Delta t \frac{\partial^2 P}{\partial x \partial t} + r\cos\theta \Delta t \frac{\partial^2 P}{\partial y \partial t} \right\} f_\Theta(\theta) d\theta$$

$$= P - \Delta t \frac{\partial P}{\partial t} + \frac{r^2}{4} \left\{ \frac{\partial^2 P}{\partial x^2} + \frac{\partial^2 P}{\partial y^2} \right\} + o(\{\Delta t\}^2)$$

From this we obtain

$$\Delta t \frac{\partial P}{\partial t} = \frac{r^2}{4} \left\{ \frac{\partial^2 P}{\partial x^2} + \frac{\partial^2 P}{\partial y^2} \right\} + o(\{\Delta t\}^2) \Rightarrow \frac{\partial P}{\partial t} = \frac{r^2}{4\Delta t} \left\{ \frac{\partial^2 P}{\partial x^2} + \frac{\partial^2 P}{\partial y^2} \right\} + o(\Delta t)$$

If we define

$$D_4 = \lim_{r, \Delta t \to 0} \left\{ \frac{r^2}{4\Delta t} \right\} \tag{7.29}$$

then in the limit as $r, \Delta t \to 0$, we obtain

$$\frac{\partial P}{\partial t} = D_4 \left\{ \frac{\partial^2 P}{\partial x^2} + \frac{\partial^2 P}{\partial y^2} \right\} \tag{7.30}$$

Thus, the result is similar to that of the two-dimensional symmetric random walk, the difference being the diffusion constant.

7.9 SUMMARY

Diffusion is a transport phenomenon that exists whenever there is a concentration gradient of some substance. It acts to eliminate the gradient so that when it is over the concentration gradient is 0 and thereafter particle motion becomes a random process that enables motion in any direction. The diffusion process is used to model the movement of objects through a medium. The object can be as small as a molecule and as large as animals and humans. The fact that a diffusion process is described by a partial differential equation enables it to be adopted in approximating many discrete processes that are usually described by differential-difference equations. In this chapter, we have discussed different diffusion processes, including the Brownian motion, and the diffusion approximation of different types of random walk. A type of diffusion called anomalous or fractional diffusion is discussed in Chapter 9.

CHAPTER 8

LEVY WALK

8.1 INTRODUCTION

Both the Poisson process and the Brownian motion have stationary and independent increments. However, they have different sample paths: the Brownian motion has continuous sample paths while the Poisson process has discontinuities (or jumps) of size 1. At a high level, Levy processes are stochastic processes with both stationary and independent increments. They constitute a wide class of stochastic processes whose sample paths can be continuous, continuous with occasional discontinuities, and purely discontinuous. In this chapter, we discuss the Levy process as well as Levy flights and Levy walk. We also discuss the power law, stable distributions, and fractals that will enable us to study the Levy distribution, Levy flights, and Levy walk.

8.2 GENERALIZED CENTRAL LIMIT THEOREM

The central limit theorem describes the behavior of sums of random variables. Specifically, let $\{X_1, X_2, \ldots, X_n\}$ be a sequence of independent and identically distributed random variables with a common probability density function (PDF) $f_X(x)$. Let the random variable $Y_n = X_1 + X_2 + \cdots + X_n$ be the sum of these random variables. Then according to the central limit theorem, provided $\sigma_X^2 < \infty$, in the limit as n becomes

Elements of Random Walk and Diffusion Processes, First Edition. Oliver C. Ibe.
© 2013 John Wiley & Sons, Inc. Published 2013 by John Wiley & Sons, Inc.

large, the PDF of Y_n is normal; that is, $\lim_{n \to \infty} Y_n \sim N(\mu, \sigma^2)$, where $\mu = n\mu_X$, $\sigma^2 = n\sigma_X^2$, and μ_X is the mean of X. Thus, we have that

$$f_{Y_n}(y) = \frac{1}{\sqrt{2\pi n\sigma_X^2}} \exp\left(\frac{-[y - n\mu_X]^2}{2n\sigma_X^2}\right) \quad -\infty < y < \infty$$

The central limit theorem does not depend on the PDF or probability mass function (PMF) of the X_i and this makes the normal distribution act as a "black hole of statistics." Thus, we say that the PDF $f_X(x)$ belongs to the domain of attraction of the Gaussian distribution, if the variance σ_X^2 is finite. The requirements for the central limit theorem to be applicable are as follows:

(a) The random variables summed must be independent.
(b) All random variables must have finite mean and finite variance.
(c) No variable can make an excessively large contribution to the sum; their contributions to the sum are essentially identical.

Consider a stochastic process that can be described by the power-law PDF $f_X(x) \sim 1/x^\gamma$, where $\gamma > 1$. This is a probability distribution in which the probability of a given value of the random variable occurring falls off very slowly with increasing value of the random variable, unlike ordinary distributions where the rate of falloff is much faster. For a process with a power-law PDF, the mean, and the second moment are given respectively by

$$E[X] = \int_{-\infty}^{\infty} x f_X(x)\, dx \sim \int_{-\infty}^{\infty} x^{1-\gamma}\, dx = \left.\frac{x^{2-\gamma}}{2-\gamma}\right|_{x=-\infty}^{\infty}$$

$$E[X^2] = \int_{-\infty}^{\infty} x^2 f_X(x)\, dx \sim \int_{-\infty}^{\infty} x^{2-\gamma}\, dx = \left.\frac{x^{3-\gamma}}{3-\gamma}\right|_{x=-\infty}^{\infty}$$

From this we observe that the mean diverges when $\gamma < 2$, and the second moment diverges when $\gamma < 3$. This means that if we allow the jump sizes in a random walk to have a power-law PDF $f_X(x) \sim x^{-\gamma}$ such that $1 < \gamma < 3$, then $E[X^2] = \infty$ and the variance is also infinite. It is a common practice to write $\gamma = 1 + \alpha$. Thus, we have that $f_X(x) \sim x^{-(1+\alpha)}$, where $0 < \alpha < 2$.

The physical significance of infinite variance can be understood by recalling that variance is a measure of central tendency. When the variance is finite, values are known to be clustered around the mean; the smaller the variance, the more the values are clustered around the mean. Generally, the probability of large deviations from the mean is very small. However, in the case of infinite variance, there is no such clustering, and regardless of the scale on which measurements are made, there is no change in their central tendency; thus, all scales look the same. Physically, this means that it is difficult or impossible to put limits on the values of the random variable that one may observe. Such values can become arbitrarily large in absolute value with a much higher frequency than is the case with better-behaved distributions such as the normal distribution.

Another property of power-law distributions is that a very small number of terms dominate all the others, which means that contributions of other terms to the sum are very negligible. This is contrary to Gaussian sums where, as we stated earlier, each term contributes essentially equally to the sum. Thus, the central limit theorem cannot be applied to power-law distributions.

The generalized central limit theorem is an extension of the classical central limit theorem that was developed to deal with sums of power-law random variables whose variances are infinite. Specifically, it is an answer to the following question: Can there be a limiting distribution for the sum of an infinite number of independent and identically distributed random variables with infinite variance, which precludes their convergence to the normal distribution? The theorem states that the answer to the question is "yes." Specifically, the sum of independent and identically distributed random variables converges to a distribution that is a member of the family of random variables that have a *stable distribution*. Thus, stable distributions are a generalization of the normal distribution.

8.3 STABLE DISTRIBUTION

A random variable is defined to be stable (also called *α-stable* or *Levy stable*) if a linear combination of two independent copies of the random variable has the same distribution as the random variable. Specifically, X is defined to be stable if for any positive numbers a and b there exists a positive number c and a real number d such that

$$aX_1 + bX_2 \sim cX + d \tag{8.1}$$

where X_1 and X_2 are independent copies of X, and $u \sim v$ means that u and v are identical in distribution. If $d=0$ for all choices of a and b, then X is said to be *strictly stable*. If the distribution of X is symmetric (i.e., if X and $-X$ have the same distribution), then X is called a *symmetric* stable random variable. For any stable random variable X that satisfies Equation 8.1, there exists a number $\alpha \in [0,2]$ such that

$$c^\alpha = a^\alpha + b^\alpha \tag{8.2}$$

There are other ways to define a stable distribution. One such definition is the following. A random variable X has a stable distribution if for $n \geq 2$ there exist a positive number c_n and a real number d_n such that if the random variables $X_1, X_2, ..., X_n$ are independent copies of X then

$$X_1 + X_2 + \cdots + X_n = c_n X + d_n \tag{8.3}$$

Another definition is as follows: A random variable X is defined to have a Levy stable distribution if there are parameters $0 < \alpha \le 2$, $-1 \le \beta \le 1$, and a real value μ such that its characteristic function has the form:

$$\Phi_X(w) = E[e^{iwX}] = \begin{cases} \exp\left\{-\sigma^\alpha |w|^\alpha \left(1 - i\beta(\text{sign } w)\tan\left\{\dfrac{\pi\alpha}{2}\right\}\right) + i\mu w\right\} & \alpha \ne 1 \\[4mm] \exp\left\{-\sigma|w|\left(1 + i\beta\left(\dfrac{2}{\pi}\right)(\text{sign } w)\ln|w|\right) + i\mu w\right\} & \alpha = 1 \end{cases}$$

$$(8.4)$$

The parameter α is called the *stability index* and

$$\text{sign } w = \begin{cases} 1 & \text{if } w > 0 \\ 0 & \text{if } w = 0 \\ -1 & \text{if } w < 0 \end{cases}$$

Stable distributions are generally characterized by four parameters:

- A *stability index* $\alpha \in [0, 2]$, which is also called the *tail index, tail exponent,* or *characteristic exponent,* determines the rate at which the tails of the distribution taper off. In particular, the smaller the value of α, the greater the frequency and size of the extreme events. When $\alpha > 1$, the mean of the distribution exists and is equal to μ.
- A *skewness parameter* $\beta \in [-1, 1]$, which has the following property: When it is positive, the distribution is skewed to the right, which means that the right tail is thicker; and when it is negative, it is skewed to the left. When $\beta = 0$, the distribution is symmetric about μ. As α approaches 2, β loses its effect and the distribution approaches the Gaussian distribution regardless of β.
- A *scale parameter* $\sigma > 0$, which determines the width and thus dispersion of the PDF.
- A *location* or *shift parameter* $\mu \in \mathfrak{R}$, which measures the shift of the mode (i.e., the peak) of the distribution, and plays the role that the mean plays in a normal distribution.

Thus, if X is a stable random variable, it is generally expressed as $X \sim S(\alpha, \beta, \sigma, \mu)$. One major drawback of the stable distribution is that, with the exception of three special cases, its PDF and cumulative distribution function (CDF) do not have closed-form expressions. Thus, the stable distribution is usually described by its characteristic function. The PDF of X is generally written in the form $f_X(x; \alpha, \beta, \sigma, \mu)$. The stability index and the skewness parameter play a more important role than the scale and shift parameters. The three special cases are as follows:

(a) The Gaussian distribution in which $\alpha=2$ and the mean is μ. The skewness parameter has no effect and so we use $\beta=0$. Thus, the PDF is $f_X(x; 2, 0, \sigma, \mu)$ and the variance is given by $\sigma_X^2 = 2\sigma^2$. The PDF and CDF are given by

$$f_X(x; 2, 0, \sigma, \mu) = \frac{1}{\sqrt{4\pi\sigma^2}} \exp\left\{-\frac{(x-\mu)^2}{4\sigma^2}\right\} \qquad -\infty < x < \infty$$

$$F_X(x; 2, 0, \sigma, \mu) = \Phi\left(\frac{x-\mu}{\sigma\sqrt{2}}\right)$$

(8.5)

(b) The Cauchy distribution in which $\alpha=1$ and $\beta=0$. Thus, the PDF and CDF are

$$f_X\left(x; 1, 0, \sigma\mu\right) = \frac{1}{\pi\sigma\left[1+\left(\dfrac{x-\mu}{\sigma}\right)^2\right]} = \frac{\sigma}{\pi\left[\sigma^2 + (x-\mu)^2\right]} \qquad -\infty < x < \infty$$

(8.6a)

$$F_X\left(x; 1, 0, \sigma, \mu\right) = \frac{1}{2} + \frac{1}{\pi}\arctan\left(\frac{x-\mu}{\sigma}\right)$$

(8.6b)

The case where $\mu=0$ and $\sigma=1$ is called the standard Cauchy distribution whose PDF and CDF become

$$f_X\left(x; 1, 0, 1, 0\right) = f_X(x; 1) = \frac{1}{\pi[1+x^2]} \qquad -\infty < x < \infty$$

(8.7a)

$$F_X\left(x; 1, 0, 1, 0\right) = \frac{1}{2} + \frac{1}{\pi}\arctan(x)$$

(8.7b)

From this we observe that the general Cauchy distribution is related to the standard Cauchy distribution as follows:

$$f_X(x; 1, 0, \sigma, \mu) = \frac{1}{\sigma}f_X\left(\frac{x-\mu}{\sigma}; 1\right)$$

(c) The Levy distribution in which $\alpha=0.5$ and $\beta=1$. Thus, the PDF and CDF are

$$f_X\left(x; \tfrac{1}{2}, 1, \sigma, \mu\right) = \left(\frac{\sigma}{2\pi}\right)^{1/2}\frac{1}{(x-\mu)^{3/2}}\exp\left\{-\frac{\sigma}{2(x-\mu)}\right\} \qquad \mu < x < \infty$$

$$F_X\left(x; \tfrac{1}{2}, 1, \sigma, \mu\right) = 2\left\{1-\Phi\left(\frac{\sigma}{x-\mu}\right)\right\}$$

(8.8)

It must be emphasized that with the exception of the Gaussian distribution (i.e., $\alpha=2$), all stable distributions are *leptokurtic* and heavy-tailed distributions. Leptokurtic distributions have higher peaks around the mean compared to the normal distribution, which leads to thick tails on both sides.

We can obtain the parameters of the equation $aX_1+bX_2=cX+d$ for the normal random variable $X \sim N(\mu_X,\sigma_X^2)$ as follows. For a normal random variable, $\alpha=2$. Since X_1 and X_2 are two independent copies of X we have that

$$E[aX_1+bX_2]=(a+b)\mu_X, \quad \mathrm{Var}(aX_1+bX_2)=\left(a^2+b^2\right)\sigma_X^2$$
$$E[cX+d]=c\mu_X+d, \quad \mathrm{Var}(cX+d)=c^2\sigma_X^2$$

Thus, we have that

$$aX_1+bX_2 \sim N\left(\{a+b\}\mu_X,\{a^2+b^2\}\sigma_X^2\right)$$
$$cX+d \sim N\left(c\mu_X+d,c^2\sigma_X^2\right)$$

Since the two are equal in distribution, we have that

$$c^2=a^2+b^2 \Rightarrow c=\sqrt{a^2+b^2}$$
$$d=(a+b-c)\mu_X$$

Note that $c^2=a^2+b^2$, which is a property of the stable distribution that we discussed earlier; that is, $c^\alpha=a^\alpha+b^\alpha$ where $\alpha=2$.

Restatement of the Generalized Central Limit Theorem: In light of our discussion on stable distributions, we can restate the generalized central limit theorem as follows. Let $X_1, X_2, \ldots, X_n$ be a sequence of independent and identically distributed random variables with a common PDF $f_X(x)$. Let $S_n=X_1+X_2+\cdots+X_n$. Let Z_α be an α-stable distribution. The random variable X belongs to the domain of attraction for Z_α (that is, PDF of an infinite sum of X_i converges to the PDF of Z_α) if there exist constants $b_n \in \mathfrak{R}$ such that

$$\frac{S_n-b_n}{n^{1/\alpha}} \sim Z_\alpha$$

That is, $(S_n-b_n)/n^{1/\alpha}$ converges in distribution to Z_α.

Figure 8.1 shows the plots of the PDFs of the three α-stable distributions for various α, β, σ, and μ values.

Stable laws are of interest for their asymptotic behavior. The asymptotic behavior of the stable distributions for $\alpha<2$ and skewness parameter β as $x \to \infty$ is of the form:

$$\lim_{x \to \pm\infty} f_X(x;\alpha,\beta,\sigma,\mu) \sim \frac{\alpha(1+\beta)C(\alpha)}{|x|^{1+\alpha}}$$

where

$$C(\alpha)=\frac{1}{\pi}\Gamma(\alpha)\sin\left(\frac{\pi\alpha}{2}\right)$$

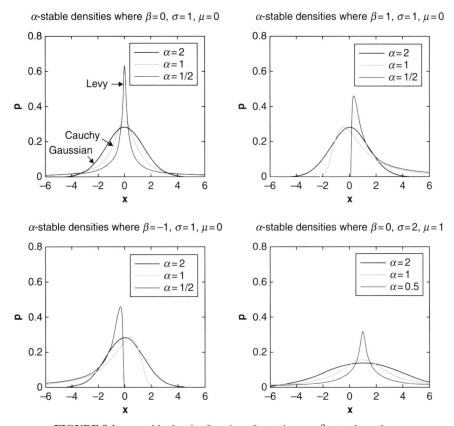

FIGURE 8.1 α-stable density functions for various α, β, σ and μ values.

$\Gamma(u)$ is the gamma function of u. Thus, for the Cauchy distribution, $C(\alpha)=C(1)=1/\pi$, for the Levy distribution, $C(\alpha) = C(0.5) = (1/\sqrt{2\pi})$; and for the normal distribution, $C(\alpha)=C(2)=0$ which is consistent with the fact that the PDF of the normal distribution goes to 0 as $x \to \infty$.

Recall that the characteristic function of the PDF $f_X(x; \alpha, \beta, \sigma, \mu)$ has the form

$$\Phi_X(w) = E[e^{iwX}] = \int_{-\infty}^{\infty} f_X(x) e^{iwx} dx$$

$$= \begin{cases} \exp\left\{-\sigma^\alpha |w|^\alpha \left(1 - i\beta(\text{sign } w) \tan\left\{\frac{\pi\alpha}{2}\right\}\right) + i\mu w\right\} & \alpha \neq 1 \\ \exp\left\{-\sigma|w|\left(1 + i\beta\left(\frac{2}{\pi}\right)(\text{sign } w) \ln|w|\right) + i\mu w\right\} & \alpha = 1 \end{cases}$$

$$f_X(x; \alpha) = \frac{1}{2\pi}\int_{-\infty}^{\infty} \Phi_X(w) e^{-iwx} dw$$

Thus, the characteristic functions of the three special cases are as follows:

- For the normal distribution where $\alpha=2$ and $\beta=0$, the characteristic function is given by $\Phi_X(w) = \exp\{i\mu w - \sigma^2 w^2\}$
- For the Cauchy distribution where $\alpha=1$ and $\beta=0$, the characteristic function is given by $\Phi_X(w) = \exp\{i\mu w - \sigma|w|\}$
- For the Levy distribution where $\alpha=1/2$ and $\beta=1$ the characteristic function is given by $\Phi_X(w) = \exp\{i\mu w - |\sigma w|^{1/2}(1 - \text{sgn } w)\}$

8.4 SELF-SIMILARITY

From our study of geometry, when two objects have the same shape, we say that they are similar. For example, we talk of similar triangles, which are triangles that have the same shape but can be of different sizes. A geometric shape is defined to be self-similar if the same geometric structures are observed independently of the distance from which the shape is looked at. That is, self-similarity is a term used to describe an object that looks roughly the same when viewed at different degrees of magnification or different scales on a dimension (space or time).

Self-similar processes were introduced by Kolmogorov in the early 1940s, but they were brought to the attention of statisticians and others in related fields in the late 1960s and early 1970s by Mandelbrot. They are used to model the presence of long-term correlation in a data set.

More formally, a self-similar process is a stochastic process that is invariant in distribution under suitable scaling of time and space; that is, it is a process that is invariant against changes in scale or size. Specifically, the stochastic process $\{X(t), t \geq 0\}$ is said to be self-similar if for any $a>0$ there exists some $b>0$ such that $\{X(at), t \geq 0\}$ has the same distribution as $\{bX(t), t \geq 0\}$. Generally, a and b are related by $b=a^H$, where H is a parameter called the *Hurst index*. Thus, a self-similar stochastic process is one in which $\{X(at), t \geq 0\}$ has the same distribution as $\{a^H X(t), t \geq 0\}$ for all $a>0$. That is,

$$X(at) \sim a^H X(t) \tag{8.9}$$

One major property of a self-similar process $X(t)$ is that it exhibits long-range dependence when $H>1/2$, which means that if we define the autocorrelation function $\rho(n) = E[X(t_0)X(t_n)]$, then

$$\sum_{n=1}^{\infty} \rho(n) = \infty$$

When $H<1/2$, $\rho(n)$ tends to decay exponentially; that is, $\rho(n)=\beta^n$ as n tends to infinity, where $\beta<1$, which means that the process exhibits short-range dependence. In this case, $\sum_{n=1}^{\infty} \rho(n) < \infty$.

Self-similarity is an attribute of many laws of nature and is a unifying concept that underlies power laws and fractals, which are discussed next.

8.5 FRACTALS

A fractal has been defined in Mandelbrot (1982) as a rough or fragmented geometric shape that can be split into parts, each of which is (at least approximately) a reduced-size copy of the whole. The word "fractal" derives from the Latin adjective *fractus*, and verb *frangere*, to break, and is used to describe fractured or ragged shapes that possess repeating patterns when viewed at increasingly fine magnifications. This means that fractals can be described as objects that contain structures nested within one another. Thus, fractals connote fraction, fracture, and fragment. In this section, we give a brief introduction to fractals. A more detailed discussion can be found in many texts such as Addison (1997).

The definition of a fractal connotes self-similarity, and the self-similarity of a fractal can be explained using the Sierpinski gasket. A large triangle A is decomposed into three congruent triangles, each of which is exactly 1/2 the size of original triangle. That is to say, if we magnify any of the three pieces of A shown in Figure 8.2 by a factor of 2, we obtain an exact replica of A. Thus, A consists of three self-similar copies of itself, each with *magnification factor* of 2. Similarly, each of the three congruent triangles is decomposed into three congruent triangles, each of which is 1/2 the size of the parent triangle, and the procedure follows the same method.

We can look deeper into A and see further copies of A, each of which is ½ as tall and ½ as wide as A. Since each of these copies is made up of three still smaller copies, we can say that the gasket consists of 9 copies, each of which is ¼ as tall and ¼ as wide as A, or 27 copies each of which is 1/8 as tall and 1/8 as wide as A, and so on. In general, at the nth step, we can divide A into 3^n self-similar pieces each of which is 2^{-n} as tall and 2^{-n} as wide as A. Figure 8.3 shows the steps used to generate the gasket in Figure 8.2.

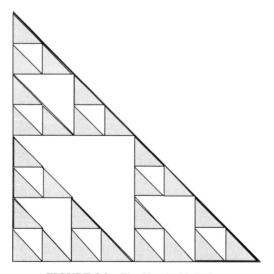

FIGURE 8.2   The Sierpinski Gasket.

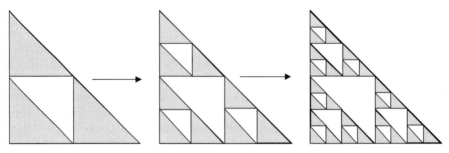

FIGURE 8.3   Steps used to generate the Sierpinski Gasket.

A fractal is said to be *self-affine* if it can be decomposed into subsets that can be linearly mapped into the full figure. A self-affine fractal is self-similar if the subsets require only rotation, translation, and dilation to map them into the full figure.

As reported in Song et al. (2006), complex networks, including the World Wide Web, the Internet, biological networks, and social networks have been shown to carry traffic that has self-repeating patterns at all length scales. Thus, these networks carry self-similar traffic.

One important parameter that characterizes a fractal is the *fractal dimension*. It is a measure of the extent to which a structure exceeds its base dimension to fill the next dimension. This means that for a fractal line, the dimension will be greater than 1 and up to 2; and for a fractal surface, the dimension will have a value between 2 and 3, and so on. One way to define fractal dimension is as follows: Consider the volume of an arbitrary object which can be measured by the number $N(l)$ of balls of linear size l. If we define the volume of a ball by l^D, where D is the dimension, then the volume of the object is

$$V = N(l)l^D$$

If we assume that the volume of the object does not change, then we have that

$$N(l) \sim l^{-D} \tag{8.10}$$

The *box-count fractal dimension* is defined as the exponent D in the relationship:

$$N(l) = l^{-D} = \frac{1}{l^D} \tag{8.11}$$

Another type of dimension is the *intrinsic dimension*, which can be defined as follows: Consider a perfectly self-similar object with r similar pieces, each of which is scaled down by a factor s. Then the fractal dimension for the object is defined by

$$D = \frac{\log(r)}{\log(s)} \tag{8.12}$$

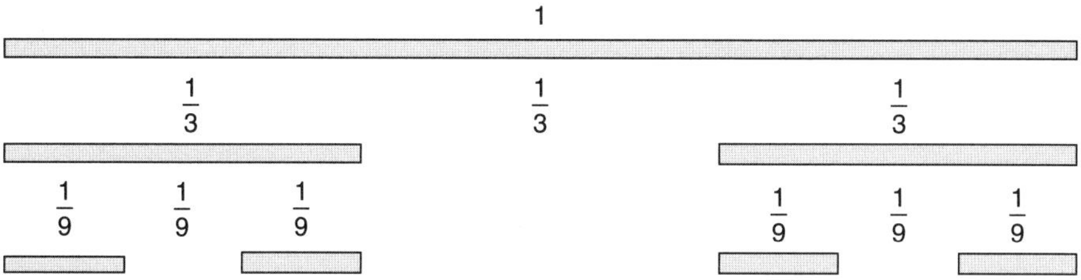

FIGURE 8.4   The Cantor Set.

For example, for the Sierpinski triangle, there are three self-similar pieces (i.e., $r=3$) each scaled down by a factor of 2 (i.e., $s=2$). Thus, the fractal dimension is

$$D = \frac{\log(3)}{\log(2)} \approx 1.58 \qquad (8.13)$$

Another example is the Cantor set, which can be constructed as follows: Start with a line segment of length 1. Then erase the middle third of the line and repeat the process for the remaining two segments of length 1/3, and so on, as shown in Figure 8.4 up to only two steps. Since the number of self-similar pieces is two, each of which is scaled down by three, the dimension of the Cantor set is

$$D = \frac{\log(2)}{\log(3)} = 0.63 \qquad (8.14)$$

   Fractal dimension analysis has been used to quantitatively assess the degree of abnormality and aggressiveness of breast cancer tumors obtained through biopsy. As reported in Tambasco et al. (2010), fractal dimension has proved to be an objective and reproducible measure of the complexity of the tissue architecture of the biopsy specimen; the higher the number, the more abnormal the tissue is.

8.6   LEVY DISTRIBUTION

The Levy distribution, named after Paul Levy, is a stable distribution with $\alpha=1/2$ and $\beta=1$. Thus, the PDF of X is given by

$$f_X\left(x; \tfrac{1}{2}, 1, \sigma, \mu\right) = \left(\frac{\sigma}{2\pi}\right)^{1/2} \frac{1}{(x-\mu)^{3/2}} \exp\left\{-\frac{\sigma}{2(x-\mu)}\right\} \quad \mu < x < \infty \quad (8.15)$$

The PDF is leptokurtic, which as we stated earlier means that it has a fat tail. This gives it the advantage over the Gaussian PDF that its fat tail accounts for a higher probability of extreme events. Both the mean and the variance are infinite. As discussed earlier, the characteristic function is given by

$$\Phi_X(w) = \exp\{i\mu w - |\sigma w|^{1/2}(1 - \operatorname{sgn} w)\} = \begin{cases} \exp\{i\mu w\} & w > 0 \\ \exp\{i\mu w - |\sigma w|^{1/2}\} & w = 0 \\ \exp\{i\mu w - 2|\sigma w|^{1/2}\} & w < 0 \end{cases}$$

Note that for very large values of x, the exponential term becomes nearly 1. This causes the PDF to be approximately $f_X(x) \sim 1/x^{3/2} = 1/x^{(1/2)+1}$ which is a function that decreases as $1/x^{3/2}$ thereby giving it fat tails.

8.7 LEVY PROCESS

As we discussed earlier, Levy processes are stochastic processes with both stationary and independent increments. They constitute a wide class of stochastic processes whose sample paths can be continuous, continuous with occasional discontinuities, and purely discontinuous. Examples of Levy processes include the Brownian motion with drift, the Poisson process, and the compound Poisson process. Among the Levy family of processes, the Brownian process with drift is the only member with continuous sample paths. All the other Levy-type processes have discontinuous trajectories and exhibit jumps.

Levy processes are widely used in the field of quantitative finance to model asset prices, and in physics. More formally, a stochastic process $\{X(t), t \geq 0\}$ is a Levy process if the following conditions are satisfied:

(a) $X(0) = 0$
(b) $X(t)$ has independent increments
(c) $X(t)$ has stationary increments
(d) $X(t)$ is continuous in probability; that is, for all $t \geq 0$ and $\varepsilon > 0$,

$$\lim_{u \to t} P\left[|X(t) - X(u)| > \varepsilon\right] = 0$$

8.8 INFINITE DIVISIBILITY

An interesting property of the Levy process is the *infinite divisibility* property. A random variable Y is said to be infinitely divisible if for every integer $n \geq 2$ there are n independent random variables $Y_{1,n}, \ldots, Y_{n,n}$ such that the sum $Y_n = Y_{1,n} + \ldots + Y_{n,n}$ has the same distribution as Y. Because the cumulative distribution function of the sum of n independent random variables is the n-way convolution of their CDFs, we have that

$$F_Y(y) = F_{Y_n}(y) = F_{Y_{1,n}}(y) * \cdots * F_{Y_{n,n}}(y) \tag{8.16}$$

In terms of the characteristic function, because the $Y_{k,n}$ are also identically distributed, we have that

$$\Phi_Y(w) = [\Phi_{Y_{1,n}}(w)]^n \tag{8.17a}$$

$$\Phi_{Y_{1,n}}(w) = [\Phi_Y(w)]^{1/n} \tag{8.17b}$$

Because for a Levy process $X(0)=0$, we can write

$$
\begin{aligned}
X(t) &= \left\{ X\left(\frac{t}{n}\right) - X\left(\frac{0t}{n}\right) \right\} + \left\{ X\left(\frac{2t}{n}\right) - X\left(\frac{t}{n}\right) \right\} + \cdots + \left\{ X\left(\frac{(n-1)t}{n}\right) - X\left(\frac{(n-2)t}{n}\right) \right\} \\
&\quad + \left\{ X\left(\frac{nt}{n}\right) - X\left(\frac{(n-1)t}{n}\right) \right\} \\
&= \sum_{k=1}^{n} \left\{ X\left(\frac{kt}{n}\right) - X\left(\frac{(k-1)t}{n}\right) \right\}
\end{aligned}
$$

Thus, $X(t)$ is the sum of n independent random variables, all of which have the same distribution as $X(t/n)$. Because the condition is true for all $n \geq 1$, we conclude that $X(t)$ has an infinitely divisible distribution.

8.8.1 The Infinite Divisibility of the Poisson Process

As stated earlier, the Poisson process is a Levy process. The characteristic function of the Poisson random variable $X(t)$ with mean λt is given by

$$
\begin{aligned}
\Phi_{X(t)}(w) &= E\left[e^{iwX(t)}\right] = \sum_{k=-\infty}^{\infty} e^{iwk} p_{X(t)}(k) = \sum_{k=0}^{\infty} e^{iwk} \left\{ \frac{e^{-\lambda t}(\lambda t)^k}{k!} \right\} \\
&= e^{-\lambda t} \sum_{k=0}^{\infty} \frac{(\lambda t e^{iw})^k}{k!} = e^{-\lambda t(1-e^{iw})} = e^{-\frac{\lambda t}{n}(1-e^{iw})n} = \left[e^{-\frac{\lambda t}{n}(1-e^{iw})} \right]^n \\
&= \left[\Phi_{Y(t)}(w) \right]^n
\end{aligned}
$$

where $Y(t)$ is a Poisson random variable with rate λ/n. Thus, the Poisson random variable has the infinite divisibility property.

8.8.2 Infinite Divisibility of the Compound Poisson Process

The compound Poisson process $X(t)$ is another example of a Levy process. Let $\Phi_Y(w)$ denote the characteristic function of the jump size density. It can be shown, using the random sum of random variable method used in Ibe (2005), that the characteristic function of the compound Poisson process is given by

$$\Phi_{X(t)}(w) = E[e^{iwX(t)}] = e^{-\lambda t(1-\Phi_Y(w))} = e^{-\frac{\lambda t}{n}(1-\Phi_Y(w))n} = \left[e^{-\frac{\lambda t}{n}(1-\Phi_Y(w))}\right]^n$$

$$= \left[\Phi_{V(t)}(w)\right]^n$$

where $V(t)$ is a Poisson random variable with rate λ/n. Thus, the compound Poisson random process has the infinite divisibility property.

8.8.3 Infinite Divisibility of the Brownian Motion with Drift

The Brownian motion with drift (or the arithmetic Brownian motion) $X(t)$ is a Levy process. We know that

$$X(t) = \mu t + \sigma B(t) \sim N(\mu t, \sigma^2 t)$$

Thus, its characteristic function is given by

$$\Phi_{X(t)}(w) = E\left[e^{iwX(t)}\right] = \int_{-\infty}^{\infty} e^{iwx} f_{X(t)}(x)dx = \int_{-\infty}^{\infty} e^{iwx} \left\{\frac{1}{\sqrt{2\pi\sigma^2 t}} e^{-\frac{(x-\mu)^2}{2\sigma^2 t}}\right\} dx$$

$$= e^{\left(i\mu wt - \frac{1}{2}\sigma^2 w^2 t\right)} = e^{\left(\frac{i\mu wt}{n} - \frac{1}{2}\frac{\sigma^2 w^2 t}{n}\right)n} = \left[e^{\left(\frac{i\mu wt}{n} - \frac{1}{2}\frac{\sigma^2 w^2 t}{n}\right)}\right]^n$$

$$= \left[\Phi_{U(t)}(w)\right]^n$$

where $\quad U(t) = \frac{\mu t}{n} + \frac{\sigma}{\sqrt{n}} B(t) \sim N\left(\frac{\mu t}{n}, \frac{\sigma^2 t}{n}\right).$

8.9 LEVY FLIGHT

A flight is the longest straight line trip that a walker takes from one location to another without a directional change or pause. A Levy flight is a mathematical description of a cluster of random short moves connected by infrequent longer ones. Thus, it consists of random walks interspersed by long travels to different regions of the walk space. In other words, there is a long movement to a random area, and then smaller movements at that area. In this way, Levy flights model activities that involve a lot of small steps that are interspersed with occasional very large excursions. For this reason it is used to model phenomena that exhibit clustering. Strictly speaking, the concept of a random walker is replaced by that of the random flyer to indicate the fact that the moving object does not touch all the intermediate lattice points but flies above them to touch down some distance from the starting point.

Levy flights have been applied to a diverse range of fields such as animal foraging patterns as in Othmer et al. (1988), Viswanathan et al. (1999), Codling (2003), Codling et al. (2008), Benhamou (2007), Edwards et al. (2007), and Buchanan (2008); the distribution of human travel as in Brockmann et al. (2006), Rhee et al. (2008), and Gonzales et al. (2008); finance and economics as in Stanley et al. (1999), Mantegna and Stanley (2000), and Porto and Roman (2002); and to model anomalous diffusion in many physical, chemical, biological and financial systems as in Metzler and Klafter (2000), Paoletti et al. (2006), Brockmann and Hufnagel (2007), Escande and Sattin (2007), Koren et al. (2007a, 2007b, 2007c), and Burnecki et al. (2008).

One of the advantages of the Levy flight over the classical (or Brownian) random walk is that the probability of a Levy flyer returning to a previously visited site is smaller than in the Brownian random walk. Also, the number of sites visited by n random walkers is much larger in the Levy flight than in the Brownian walk. Thus, n Levy flyers diffuse so rapidly that the competition for target sites among themselves is greatly reduced compared to the competition encountered by n Brownian walkers. The latter typically remain close to the origin and hence close to each other. Thus, Levy flights are superdiffusive, which implies that the variance grows faster than linearly in time with the result that the displacement from the starting point increases faster than what is predicted by a simple random walk model.

In a Levy flight, the PDF of the length X of each step is a power-law function of the form

$$f_X(x) \sim x^{-\alpha}, \quad 0 < \alpha < 2 \tag{8.18}$$

In a random walk with power-law jump sizes, higher values of α are marked by many small steps and an occasional very long movement. Thus, the few large-sized jumps the Levy flyer takes make it difficult for him to return to the starting point. If we take the logarithm of both sides of the equation, the power-law relation appears as a straight line, which implies that there is a constant proportion of steps of different lengths. Also, the second moment diverges, as we discussed earlier.

It is possible to require the flight length to be no less than some prescribed value. In this case we have that

$$f_X(x) = Cx^{-\alpha}, \quad x \geq k \tag{8.19}$$

where $k > 0$ is some constant and the normalization constant C is obtained as follows:

$$\int_k^\infty f_X(x)dx = 1 = C\int_k^\infty x^{-\alpha}dx \Rightarrow C = (\alpha - 1)k^{\alpha-}$$

Observe that for $0 < \alpha < 2$,

$$E[X] = \int_k^\infty xf_X(x)dx = C\int_k^\infty x^{1-\alpha}dx = \infty$$

$$E[X^2] = \int_k^\infty x^2 f_X(x)dx = C\int_k^\infty x^{2-\alpha}dx = \infty$$

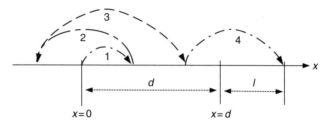

FIGURE 8.5   Illustration of first passage time.

This shows that this flight length requirement does not affect the basic properties of the process. Levy flights are closely related to fractals. Because the flight lengths in a Levy flight have a power-law distribution, the resulting pattern is a fractal; that is, it is scale-invariant and self-similar.

8.9.1   First Passage Time of Levy Flights

The first passage time (FPT) is of interest when it is required to know when a process crosses a given threshold value for the first time, such as when a share of a stock crosses a preset market value in the stock market. For example, in the gambler's ruin problem, the FPT is the time until the gambler is ruined. Consider a random flyer that starts at the origin and performs jumps whose lengths X are independent Levy stable random variables with PDF $f_X(x; \alpha)$. FPT measures how long it will take for the flyer to arrive at or cross a prescribed target. For example, in Figure 8.5, we would like to know the first time the flyer reaches or crosses the point $x=d$. From the figure we observe that it takes four jumps to cross the point $x=d$, overshooting it by a distance l. In Chechkin (2003) and Koren (2007a, 2007b, 2007c), it is shown that the asymptotic PDF of the FPT T in a symmetric Levy flight is given by

$$f_T(t;\alpha) \sim t^{-3/2}$$

$$(8.20)$$

which is consistent with the so-called Sparre Andersen result; see Sparre Andersen (1953, 1954).

8.9.2   Leapover Properties of Levy Flights

Recall that in a Levy flight the random flyer does not touch all the intermediate points but flies above them to touch down at some distance from the starting point. Therefore, in dealing with FPT in Levy flight, it is also of interest to investigate the behavior of the *first passage leapover*, which is the distance the random flyer overshoots the target in a single jump. As stated earlier, in Figure 8.5 the flyer overshot the point $x=d$ by a distance l. Let $f_L(l)$ denote the PDF of the leapover length. In Koren (2007a, 2007b, 2007c) it is shown that for a symmetric Levy flight if the asymptotic PDF of the Levy jump lengths is $f_X(x; \alpha) \sim |x|^{-1-\alpha}$, $0 < \alpha < 2$, then

$$f_L(l) \sim l^{-1-\alpha/2} \tag{8.21}$$

which implies that $f_L(l)$ is much broader than the original jump length distribution.

8.10 TRUNCATED LEVY FLIGHT

The Levy distribution has both positive and negative attributes. On the positive side, the scaling property is a very attractive feature, which says that the PDF of the sum of N Levy random variables is, up to a certain scaling factor, identical to that of the individual random variables. This is a typical fractal behavior that says that the whole looks like its parts. Thus, if $V = X_1 + X_2 + \cdots + X_N$, where the X_i are Levy random variables, then we have that

$$f_V(v) = \frac{1}{N^{1/\alpha}} f_X\left(\frac{v}{N^{1/\alpha}}\right)$$

On the negative side, the Levy distribution has divergent mean and variance, which makes it impossible to use the central limit theorem.

The truncated Levy flight (TLF) is a method introduced by Mantegna and Stanley (1994) to deal with the infinite variance associated with the Levy distribution. A finite variance is induced by truncating the tails of the Levy distribution. If $f_X(x)$ is the PDF of the Levy random variable, the PDF associated with the TLF is defined by

$$f_Y(y) = \begin{cases} cf_X(y) & -y_0 \leq y \leq y_0 \\ 0 & \text{otherwise} \end{cases} \tag{8.22}$$

where c is a normalizing constant and y_0 is the cutoff parameter. The TLF has a finite variance but retains the power-tail property of the Levy distribution within a large but finite interval. Since its variance is finite, the distribution of the sum of a large number of TLF variables will converge to a normal distribution, in the spirit of the central limit theorem. The value of y_0 has been shown through simulation to have a great influence on the TLF. If y_0 is large, $f_Y(y)$ has a pronounced Levy distribution character that drastically retards the convergence to a normal distribution when summing random variables with TLF distribution.

8.11 LEVY WALK

A Levy walk is similar to a Levy flight. The difference between the two lies in the concept of velocity. In a Levy flight, a walker visits only the endpoints of a jump and the notion of velocity does not arise; alternatively, jumps take zero or a vanishingly small time to complete. Put in another way, the velocity in the process is infinity and

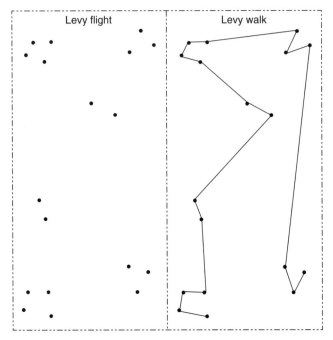

FIGURE 8.6 Difference between Levy flight and Levy walk.

it takes a *constant time* for every flight irrespective of the length of the flight. Thus, for a single jump in a Levy flight, the walker is only at the starting point and endpoint and never in any place between the two. For a single jump in a Levy walk, the walker follows a continuous trajectory from the starting point to the endpoint and hence a finite time is needed to complete the walk. For this reason, the velocity is finite and a Levy walk takes a flight time that is proportional to the flight length. Basically, the sites visited by a Levy flight are the turning points of a Levy walk. Figure 8.6 illustrates the difference between Levy flight and Levy walk.

Another way to discuss the distinction between a Levy walk and a Levy flight is based on their mobility speed. Levy flight is a "fast" mobility model while a Levy walk is a "slow" mobility model. The mean square displacement for a jump in a Levy flight is infinite while in the Levy walk where the walker moves with a finite velocity the mean square displacement is finite, but it is a time-dependent quantity.

Levy flights are unrealistic in many physical systems because physically realizable systems cannot have diverging moments. On the other hand, Levy walks do not violate any physical laws since no physically measurable quantity diverges for a Levy walk in finite time.

8.11.1 Levy Walk as a Coupled CTRW

Recall that in our discussion on the continuous-time random walk (CTRW) in Chapter 3, we defined T as the waiting time, X as the jump size, $f_X(x)$ as the PDF of X, and $f_T(t)$ as the PDF of T. The probability $P(x, t)$ that the position of the walker at time t is x, given that it was in position 0 at time $t=0$, is given by the master equation for the uncoupled CTRW as follows:

$$P(x, t) = \delta(x)R_T(t) + \int_0^t \int_{-\infty}^{\infty} P(u, \tau)f_X(x-u)f_T(t-\tau)du\,d\tau$$

where $R_T(t)=P[T>t]$. Also, if the joint Fourier–Laplace transform of $P(x, t)$ is $P^*(w, s)$, then the master equation is transformed into

$$P^*(w, s) = \int_0^{\infty} e^{-st} \int_{-\infty}^{\infty} e^{jwx} \left[\delta(x)R_T(t) + \int_0^t \int_{-\infty}^{\infty} P(u, \tau)f_X(x-u)f_T(t-\tau)du\,d\tau \right] dx\,dt$$

$$= \tilde{R}_T(s) + P^*(w, s)\Phi_X(w)\Phi_T(s)$$

where $\tilde{R}_T(s)$ is the Laplace transform of $R(t)$, $\Phi_X(w)$ is the Fourier transform of $f_X(x)$ and $\Phi_T(s)$ is the Laplace transform (or s-transform) of $f_T(t)$. This gives

$$P^*(w, s) = \frac{\tilde{R}_T(s)}{1 - \Phi_X(w)\Phi_T(w)}$$

Since

$$\tilde{R}_T(s) = \frac{1 - \Phi_T(s)}{s},$$

we have that

$$P^*(w, s) = \frac{\tilde{R}_T(s)}{1 - \Phi_X(w)\Phi_T(s)} = \frac{1 - F_T(s)}{s\left[1 - \Phi_X(w)\Phi_T(s)\right]}$$

As stated earlier, the difference between the Levy flight and the Levy walk is the velocity of propagation. In the Levy flight, it is assumed that a jump takes zero or vanishingly small time, while in the Levy flight, it proceeds at a constant nonzero velocity. This means that in Levy walk the distance and time are coupled with the result that the process can be modeled by a coupled CTRW. Thus, we may assume that the PDFs of X and T are given respectively as follows:

$$f_X(x) \sim |x|^{-(1+\alpha)} \quad 0 < \alpha < 2$$
$$f_T(t) \sim t^{-(1+\beta)} \quad \beta > 0$$

From these we obtain

$$\Phi_X(w) \sim 1 - |w|^\alpha + \cdots$$
$$\Phi_T(s) \sim 1 - s^\beta + \cdots$$

For a Levy walk where X and T are coupled, we proceed as follows. Let $Q(x, t)$ be the probability that the walker just arrived at position x at time t. Then the master equation becomes

$$Q(x, t) = \delta(x)\delta(t) + \int_0^t \int_{-\hbar}^{\infty} Q(u, \tau) f_{XT}(x - u, t - \tau) du \, d\tau \tag{8.23}$$

The joint Fourier–Laplace transform of the preceding equation is given by

$$Q^*(w, s) = 1 + Q^*(w, s)\Phi_{XT}(w, s)$$

which gives

$$Q^*(w, s) = \frac{1}{1 - \Phi_{XT}(w, s)} \tag{8.24}$$

Let v denote the velocity of propagation. Then we may write

$$f_{XT}(x, t) = f_X(x) f_{T|X}(t \mid x) = f_X(x)\delta(x - vt)$$

which ensures that $|x| = vt$. The Fourier–Laplace transform of $f_{XT}(x, t)$ can be obtained as follows. Let $\Phi_{XT}(w, t)$ denote the Fourier transform of $f_{XT}(x, t)$; that is,

$$\Phi_{XT}(w, t) = \int_{-\infty}^{\infty} f_{XT}(x, t) e^{-jwx} dx = \int_{-\infty}^{\infty} f_X(x)\delta(x - vt) e^{-jwx} dx$$
$$= f_X(vt) e^{-jwvt}$$

The Laplace transform of $\Phi_{XT}(w, t)$ is given by

$$\Phi_{XT}(w, s) = \int_0^{\infty} \Phi_{XT}(w, t) e^{-st} dt = \int_0^{\infty} f_X(vt) e^{-jwvt} e^{-st} dt = \int_0^{\infty} f_X(vt) e^{-t(s + jwv)} dt$$
$$= \frac{1}{v} \Phi_X\left(\frac{s + jwv}{v}\right) = \frac{1}{v}\Phi_X\left(\frac{s}{v} + jw\right) \tag{8.25}$$

where we note that the second to the last equation arises from the fact that the integral is the Laplace transform of $e^{-jwvt} f_X(vt)$ and we know that the Laplace transform of $f_X(vt)$ is $\dfrac{1}{v}\Phi_X\left(\dfrac{s}{v}\right)$. Thus, if we assume that the jump size distribution follows a power law as before, we may write

$$f_X(x) \sim |x|^{-(1+\alpha)}, \quad 0 < \alpha < 2$$

Note also that we can obtain the same result by finding the Laplace transform of $f_{XT}(x, t)$ followed by the Fourier transform of the result as follows:

$$\Phi_{XT}(x, s) = \int_0^\infty f_{XT}(x, t) e^{-st} dt = \int_0^\infty f_X(x) \delta(x - vt) e^{-st} dt = f_X(x) \int_0^\infty \delta(x - vt) e^{-st} dt$$

$$= \frac{1}{v} f_X(x) e^{-sx/v}$$

where the last equality follows from the fact that $\delta(x - vt) = \delta(v[(x/v) - t]) = (1/v)\delta((x/v) - t)$. Thus, taking the Fourier transform we obtain

$$\Phi_{XT}(w, s) = \int_{-\infty}^\infty \Phi_{XT}(x, s) e^{-jwx} dx = \int_{-\infty}^\infty \frac{1}{v} f_X(x) e^{-sx/v} e^{-jwx} dx = \frac{1}{v} \int_{-\infty}^\infty f_X(x) e^{-x(s/v + jw)} dx$$

$$= \frac{1}{v} \Phi_X\left(\frac{s}{v} + jw\right)$$

as before.

8.11.2 Truncated Levy Walk

As in the case of the truncated Levy flight, we can define a truncated Levy walk (TLW) whose flight lengths and wait times have truncated power-law distributions; that is,

- Flight lengths follow a truncated power law: $f_R(r) \sim |r|^{-(1+\alpha)}, r < r_c$
- Pause times follow a truncated power law: $f_T(t) \sim |t|^{-(1+\gamma)}, t < t_c$

Also, the turning angles follow a uniform distribution; that is, $f_\Theta(\theta) = 1/2\pi$, for $0 \le \theta \le 2\pi$.

8.12 SUMMARY

This chapter has discussed the Levy flight and the Levy walk and many of the concepts associated with them including power-law distribution, self-similarity, fractals, stable distribution, and the generalized central limit theorem. The latter is an extension of the central limit theorem and is defined for random variables that have a power-law distribution and thus have infinite variance. The chapter also discusses the difference between the Levy flight and the Levy walk. In a Levy flight, a walker visits only the endpoints of a jump and no point in between. In a Levy walk, the walker follows a continuous trajectory from the starting point to the endpoint and hence a finite time is needed to complete the walk. Thus, the sites visited by a Levy flight are the turning points of a Levy walk. The truncated versions of both the Levy walk and Levy flight are also discussed.

CHAPTER 9

FRACTIONAL CALCULUS AND ITS APPLICATIONS

9.1 INTRODUCTION

For the function $f(x)$, the notations $\dfrac{df(x)}{dx} \equiv D^1 f(x)$ and $\dfrac{d^2 f(x)}{dx^2} \equiv D^2 f(x)$ are well-understood to mean the first derivative and second derivative, respectively. But a notation like $\dfrac{d^{1/2} f(x)}{dx^{1/2}}$ or $D^{1/2}f(x)$ is not familiar to many people. This is what fractional calculus deals with. That is, fractional calculus is the branch of calculus that generalizes ordinary calculus by letting differentiation and integration be of any arbitrary, real, or complex order. It has become popular due to its demonstrated applications in many diverse fields of science and engineering.

The increase in its use in applications arises from the fact that it has been observed that while many analytical models assume that there is no memory in the system under investigation, many of these systems actually exhibit some form of memory. Fractional calculus has proved to be a natural tool for modeling memory-dependent phenomena, which accounts for its increasing popularity in memory-dependent systems modeling.

The concept of fractional calculus is believed to have stemmed from a question that L'Hôpital asked Leibniz in 1695 about the meaning of the derivative of order n, $d^n y/dx^n$, when $n = 1/2$. Leibniz is said to have responded as follows: "This is an apparent paradox from which, one day, useful consequences will be drawn." Today,

Elements of Random Walk and Diffusion Processes, First Edition. Oliver C. Ibe.
© 2013 John Wiley & Sons, Inc. Published 2013 by John Wiley & Sons, Inc.

fractional calculus has found application in such areas as fluid flow, dynamic processes in self-similar and porous structures, diffusive transport, electrical networks, chemical physics, and signal processing. In this chapter, we review the basics of fractional calculus and discuss some of its applications. We will use the term "classical calculus" to refer to the traditional (or ordinary) calculus. More detailed discussion on fractional calculus can be found in such texts as Oldham and Spanier (1974), Podlubny (1999), Kilbas et al. (2006), Diethelm (2010), and Herrmann (2011).

9.2 GAMMA FUNCTION

The factorial of an integer n, denoted by $n!$, is defined by

$$n! = n \times (n-1) \times \cdots \times 3 \times 2 \times 1 = \prod_{k=1}^{n} k \tag{9.1}$$

Its properties include the following:

$$1! = 1$$
$$n! = n(n-1)!$$

By definition, $0! = 1$. The gamma function, which was introduced by Euler, is a generalization of the factorial to noninteger values. For any $\alpha > 0$, the gamma function is defined by

$$\Gamma(\alpha) = \int_{0}^{\infty} e^{-t} t^{\alpha-1} dt = 2 \int_{0}^{\infty} e^{-t^2} t^{2\alpha-1} dt \tag{9.2}$$

It can be shown that

$$\Gamma(\alpha + 1) = \alpha \Gamma(\alpha).$$

Also, when α is an integer,

$$\Gamma(\alpha) = (\alpha - 1)!$$

It can also be shown that

$$\Gamma\left(\frac{1}{2}\right) = \sqrt{\pi}$$

Thus, since $\Gamma(\alpha+1) = \alpha\Gamma(\alpha)$, we have that

$$\Gamma\left(\tfrac{3}{2}\right) = \tfrac{1}{2}\Gamma\left(\tfrac{1}{2}\right) = \tfrac{1}{2}\sqrt{\pi}$$
$$\Gamma\left(\tfrac{5}{2}\right) = \tfrac{3}{2}\Gamma\left(\tfrac{3}{2}\right) = \tfrac{3}{4}\sqrt{\pi}$$

The gamma function is related to the *beta function* $B(x,y)$, $x>0$, $y>0$, as follows:

$$B(x,y) = \frac{\Gamma(x)\Gamma(y)}{\Gamma(x+y)} \tag{9.3}$$

where

$$B(x,y) = \int_0^1 u^{x-1}(1-u)^{y-1}\,du \tag{9.4}$$

9.3   MITTAG–LEFFLER FUNCTIONS

The exponential function e^x is defined as follows:

$$e^x = \sum_{n=0}^{\infty} \frac{x^n}{n!} = \sum_{n=0}^{\infty} \frac{x^n}{\Gamma(n+1)} \tag{9.5}$$

The Mittag–Leffler function is an extension of the exponential function to arbitrary real numbers $\alpha > 0$, and defined as follows:

$$E_\alpha(x) = \sum_{n=0}^{\infty} \frac{x^n}{\Gamma(n\alpha+1)} \tag{9.6}$$

The Mittag–Leffler function for special values for some integer values of α are as follows:

$$E_0(x) = \sum_{n=0}^{\infty} \frac{x^n}{\Gamma(1)} = \sum_{n=0}^{\infty} x^n = \frac{1}{1-x}$$

$$E_1(x) = \sum_{n=0}^{\infty} \frac{x^n}{\Gamma(n+1)} = e^x$$

$$E_2(x) = \sum_{n=0}^{\infty} \frac{x^n}{\Gamma(2n+1)} = \cosh\left(\sqrt{x}\right)$$

The more general Mittag–Leffler function can be defined for α, $\beta > 0$ as follows:

$$E_{\alpha,\beta}(x) = \sum_{n=0}^{\infty} \frac{x^n}{\Gamma(n\alpha + \beta)} \qquad (9.7)$$

This means that

$$E_{\alpha}(x) = E_{\alpha,1}(x) \qquad (9.8)$$

The following are some of the properties of $E_{\alpha,\beta}(x)$:

1. $E_{\alpha,\beta}(x) = xE_{\alpha,\alpha+\beta}(x) + \dfrac{1}{\Gamma(\beta)}$

2. $E_{\alpha,\alpha+\beta}(x) = \dfrac{1}{x}\left\{ E_{\alpha,\beta}(x) - \dfrac{1}{\Gamma(\beta)} \right\}$

3. $E_{\alpha,\beta}(x) = \beta E_{\alpha,\beta+1}(x) + \alpha x \dfrac{d}{dx} E_{\alpha,\beta+1}(x)$

4. $E_{\alpha,\beta}^{(r)}(x) = \dfrac{d^r}{dx^r} E_{\alpha,\beta}(x) = \displaystyle\sum_{k=0}^{\infty} \frac{(k+r)!\,x^k}{k!\,\Gamma(k\alpha + r\alpha + \beta)} = r!\displaystyle\sum_{k=r}^{\infty}\binom{k}{r}\frac{x^{k-r}}{\Gamma(k\alpha + \beta)}$,
   $r = 0,1,2,\ldots$

5. $E_{\alpha,\beta}'(x) = \dfrac{d}{dx} E_{\alpha,\beta}(x) = \dfrac{E_{\alpha,\beta-1}(x) - (\beta-1)E_{\alpha,\beta}(x)}{\alpha x}$, where $\beta \neq 1$

6. $\dfrac{d^m}{dx^m}\left\{ x^{\beta-1} E_{\alpha,\beta}(x^\alpha) \right\} = x^{\beta-m-1} E_{\alpha,\beta-m}(x^\alpha)$, $\quad \mathrm{Re}(\beta - m) > 0$, $\quad m = 0,1,\ldots$

7. $\displaystyle\int_0^t E_{\alpha,\beta}(\lambda x^\alpha) x^{\beta-1}\,dx = t^\beta E_{\alpha,\beta+1}(\lambda t^\alpha)$, $\quad \beta > 0$, which can be obtained by integrating Equation 9.7 term by term.

Examples of simple functions that can be expressed in terms of the generalized Mittag–Leffler function include:

$$E_{1,2}(x) = \sum_{n=0}^{\infty} \frac{x^n}{\Gamma(n+2)} = \sum_{n=0}^{\infty} \frac{x^n}{(n+1)!} = \frac{1}{x}\sum_{n=0}^{\infty}\frac{x^{n+1}}{(n+1)!} = \frac{e^x - 1}{x}$$

$$E_{1,3}(x) = \sum_{n=0}^{\infty} \frac{x^n}{\Gamma(n+3)} = \sum_{n=0}^{\infty} \frac{x^n}{(n+2)!} = \frac{1}{x^2}\sum_{n=0}^{\infty}\frac{x^{n+2}}{(n+2)!} = \frac{e^x - 1 - x}{x^2}$$

In general,

$$E_{1,m}(x) = \frac{1}{x^{m-1}}\left\{ e^x - \sum_{n=0}^{m-2} \frac{x^n}{\Gamma(n+1)} \right\}$$

Also,

$$E_{2,1}\left(-x^2\right) = \sum_{n=0}^{\infty} \frac{(-1)^n x^{2n}}{\Gamma(2n+1)} = \sum_{n=0}^{\infty} \frac{(-1)^n x^{2n}}{(2n)!} = \cos(x)$$

$$E_{2,2}\left(-x^2\right) = \sum_{n=0}^{\infty} \frac{(-1)^n x^{2n}}{\Gamma(2n+2)} = \frac{1}{x}\sum_{n=0}^{\infty} \frac{(-1)^n x^{2n+1}}{(2n+1)!} = \frac{\sin(x)}{x}$$

9.4   LAPLACE TRANSFORM

The Laplace transform of a function $f(t)$, denoted by $L\{f(t)\}$ or $F(s)$, is defined by the following equation:

$$F(s) = L\{f(t)\} = \int_0^{\infty} e^{-st} f(t)dt \tag{9.9}$$

The inverse Laplace transform enables us to recover $f(t)$, and is given by

$$f(t) = L^{-1}\{F(s)\} \tag{9.10}$$

The Laplace transform is commonly used in the solution of differential equations. Some of its properties include the following:

1. *Linearity property*: If $f(t)$ and $g(t)$ are two functions and a and b are two real numbers, then $L\{af(t)+bg(t)\} = aF(s)+bG(s)$
2. *Frequency differentiation*: $L\{t^n f(t)\} = (-1)^n F^{(n)}(s)$, where $F^{(n)}(s)$ is the nth derivative of $F(s)$.
3. *Differentiation*: If $f(t)$ is n times differentiable, then

$$L\left\{\frac{d^n f(t)}{dt^n}\right\} = L\left\{f^{(n)}(t)\right\} = s^n F(s) - \sum_{k=0}^{n-1} s^{n-k-1} f^{(k)}(0)$$

$$= s^n F(s) - \sum_{k=0}^{n-1} s^k f^{(n-k-1)}(0) \tag{9.11}$$

In particular,

$$L\left\{\frac{df(t)}{dt}\right\} = L\{f'(t)\} = sF(s) - f(0)$$

4. *Integration*:

a. $L\left\{\int_0^t f(u)du\right\} = \dfrac{1}{s}F(s) = s^{-1}F(s)$

b. $L\left\{\int_0^t \cdots \int_0^t f(u)du^n\right\} = L\left\{\int_0^t \dfrac{(t-u)^{n-1}}{(n-1)!}f(u)du\right\} = \dfrac{1}{s^n}F(s) = s^{-n}F(s)$

5. *Time Scaling*: $L\{f(at)\} = \dfrac{1}{|a|}F\left(\dfrac{s}{a}\right)$

6. *Convolution*: $L\{f(t)*g(t)\} = F(s)G(s)$
7. *Frequency shift*: $L\{e^{at}f(t)\} = F(s-a)$

8. *Frequency integration*: $L\left\{\dfrac{f(t)}{t}\right\} = \int_s^\infty F(u)du$

9. *Constant term*: $L\{a\} = \dfrac{a}{s}$

10. *Time shifting*: For $t \ge a$, $L\{f(t-a)\} = e^{-as}F(s)$

11. *Power*: $L\{t^n\} = \dfrac{n!}{s^{n+1}}, \quad n = 0, 1, 2, \ldots$

12. *Mittag–Leffler functions*:

a. $L\left\{x^{\beta-1}E_{\alpha,\beta}(bx^\alpha)\right\} = \dfrac{s^{\alpha-\beta}}{s^\alpha - b}$

b. $L\left\{x^{\beta-1}E_{\alpha,\beta}(-bx^\alpha)\right\} = \dfrac{s^{\alpha-\beta}}{s^\alpha + b}$

c. $L\left\{E_\alpha(bx^\alpha)\right\} = \dfrac{s^{\alpha-1}}{s^\alpha - b}$

d. $L\left\{E_\alpha(-bx^\alpha)\right\} = \dfrac{s^{\alpha-1}}{s^\alpha + b}$

e. $L\left\{x^{\alpha-1}E_{\alpha,\alpha}(-bx^\alpha)\right\} = \dfrac{1}{s^\alpha + b}$

f. $L\left\{x^{\alpha k}E_\alpha^{(k)}(-x^\alpha)\right\} = L\left\{x^{\alpha k}\dfrac{d^k}{dx^k}E(-x^\varepsilon)\right\} = \dfrac{k!s^{\alpha-1}}{(s^\alpha + 1)^{k+1}} \quad k = 0, 1, 2, \ldots$

9.5   FRACTIONAL DERIVATIVES

Consider the derivatives of $f(x)=x^k$, $k \in N$, in classical calculus. The nth derivative of $f(x)$, $n \in N$, is given by

$$\frac{d^n f}{dx^n} = D^n x^k = k(k-1)\cdots(k-n-1)x^{k-n} = \frac{k!}{(k-n)!}x^{k-n}$$

$$= \frac{\Gamma(k+1)}{\Gamma(k-n+1)}x^{k-n} \tag{9.12}$$

We can extend the derivatives of $f(x)=x^k$ to the case when the positive integer n is replaced by the arbitrary number α in terms of the gamma function as follows:

$$\frac{d^\alpha f}{dx^\alpha} = D^\alpha x^k = \frac{\Gamma(k+1)}{\Gamma(k-\alpha+1)}x^{k-\alpha} \quad \alpha > 0 \tag{9.13}$$

This result allows us to extend the concept of a fractional derivative to a large number of functions. We can extend the result to any function that can be expanded in a Taylor series in powers of x as follows:

$$f(x) = \sum_{k=0}^{\infty} a_k x^k$$

$$D^\alpha f(x) = \sum_{k=0}^{\infty} a_k D^\alpha x^k = \sum_{k=0}^{\infty} a_k \frac{\Gamma(k+1)}{\Gamma(k-\alpha+1)}x^{k-\alpha} \tag{9.14}$$

This result enables us to evaluate the fractional derivative of a function such as $f(x)=e^x$. From the Taylor series, we have that

$$e^x = \sum_{k=0}^{\infty}\frac{1}{k!}x^k \Rightarrow D^\alpha e^x = \sum_{k=0}^{\infty}\frac{1}{\Gamma(k+1)}\left\{\frac{\Gamma(k+1)}{\Gamma(k-\alpha+1)}x^{k-\alpha}\right\} = \sum_{k=0}^{\infty}\frac{1}{\Gamma(k-\alpha+1)}x^{k-\alpha}$$

$$\tag{9.15}$$

From Equation 9.13, we have that

$$\frac{d^{1/2}}{dx^{1/2}}x = D^{1/2}x = \frac{\Gamma(1+1)}{\Gamma\left(1-\frac{1}{2}+1\right)}x^{1-1/2} = \frac{\Gamma(2)}{\Gamma\left(\frac{3}{2}\right)}x^{1/2} = \frac{2x^{1/2}}{\sqrt{\pi}}$$

Also, the derivative of a constant, say M, is given by

$$\frac{d^\alpha}{dx^\alpha}M = {}_0D_x^\alpha Mx^0 = M\,{}_0D_x^\alpha x^0 = M\frac{\Gamma(0+1)}{\Gamma(0-\alpha+1)}x^{0-\alpha} = \frac{Mx^{-\alpha}}{\Gamma(1-\alpha)}$$

Thus, while the regular derivative of a constant is 0, the fractional derivative is nonzero.

9.6 FRACTIONAL INTEGRALS

Because integration and differentiation are inverse operations, we may write

$$D^{-1}f(x) = If(x) = \int_0^x f(u)du$$

Similarly, we have that

$$D^{-2}f(x) = \int_0^x \int_{u_1}^x f(u_1)du_2 du_1 = \int_0^x f(u_1)\int_{u_1}^x du_2 du_1 = \int_0^x f(u_1)(x - u_1)du_1$$

$$\equiv \int_0^x f(u)(x - u)du$$

$$D^{-3}f(x) = \int_0^x \int_{u_1}^x \int_{u_2}^x f(u_1)du_3 du_2 du_1 = \int_0^x f(u_1)\int_{u_1}^x \int_{u_2}^x du_3 du_2 du_1$$

$$= \int_0^x f(u_1)\int_{u_1}^x (x - u_2)du_2 du_1 = \frac{1}{2}\int_0^x f(u_1)(x - u_1)^2 du_1$$

$$\equiv \frac{1}{2}\int_0^x f(u)(x - u)^2 du$$

In the same way, it can be shown that

$$D^{-4}f(x) = \frac{1}{3 \times 2}\int_0^x f(u)(x - u)^3 du$$

In general, we have that

$$D^{-n}f(x) = \frac{1}{(n-1)!}\int_0^x f(u)(x-u)^{n-1}du = \frac{1}{\Gamma(n)}\int_0^x f(u)(x-u)^{n-1}du \quad (9.16)$$

Thus, the fractional integral of order α is defined, for a function $f(x)$, by

$$I_x^\alpha f(x) = \frac{1}{\Gamma(\alpha)}\int_0^x f(u)(x-u)^{\alpha-1}du = \frac{1}{\Gamma(\alpha)}\int_0^x \frac{f(u)}{(x-u)^{1-\alpha}}du, \ \alpha > 0 \quad (9.17)$$

9.7 DEFINITIONS OF FRACTIONAL INTEGRO-DIFFERENTIALS

There are several definitions of the fractional integro-differentials. Each of the definitions has its advantages and drawbacks and the choice depends mainly on the purpose and the area of application. The Riemann–Liouville (RL) derivative was the first fractional derivative to be developed and has a well-established mathematical theory. However, it has certain features that lead to difficulties when it is applied to real-world problems. Because of these limitations, many other models were developed. One of these models is the Caputo derivative, which has a close relationship to the RL model but with certain modifications that attempt to avoid the limitations of the RLmodel.

The three most frequently used definitions for the general fractional integro-differentials are: the Riemann–Liouville, the Grünwald–Letnikov definition, and the Caputo definition. We give a brief introduction to these definitions.

9.7.1 Riemann–Liouville Fractional Derivative

We can generalize differentiation to non-integer values by replacing the $-n$ in Equation 9.16 with arbitrary α and the factorial with the gamma function to obtain the so-called *RL derivative*:

$$D^{\alpha} f(x) = \frac{1}{\Gamma(-\alpha)} \int_0^x \frac{f(u)}{(x-u)^{\alpha+1}} \, du \tag{9.18}$$

In the preceding equation, the lower limit of integration is 0. The limits of integration are generally indicated with subscripts as follows:

$$_b D_x^{\alpha} f(x) = \frac{1}{\Gamma(-\alpha)} \int_b^x \frac{f(u)}{(x-u)^{\alpha+1}} \, du \tag{9.19}$$

That is, the left subscript is the lower limit and the right subscript is the upper limit of the integration. Similarly, we may write

$$_b D_x^{-\alpha} f(x) = {_b I_x^{\alpha}} f(x) = \frac{1}{\Gamma(\alpha)} \int_b^x f(u)(x-u)^{\alpha-1} \, du$$

$$= \frac{1}{\Gamma(\alpha)} \int_b^x \frac{f(u)}{(x-u)^{1-\alpha}} \, du, \quad \alpha > 0 \tag{9.20}$$

The following law of exponents holds whenever α and β are natural numbers:

$$D^{\alpha} D^{\beta} = D^{\alpha+\beta} \tag{9.21}$$

In particular, let $m = \lceil \alpha \rceil$; that is, m is the smallest integer that is greater than α. Also, let $v = m - \alpha$. Then

$$_0 D_x^{\alpha} f(x) = D^m \left\{ _0 D_x^{-v} f(x) \right\} \tag{9.22}$$

Thus, the fractional derivative involves integer-order differentiation of an integral. For example, let $\alpha=0.4$. Then $m=\lceil 0.4 \rceil=1$ and $v=m-\alpha=0.6$. Thus,

$$_0D_x^{0.4}f(x) = D\left\{_0D_x^{-0.6}f(x)\right\} = D\left\{\frac{1}{\Gamma(0.6)}\int_0^x (x-u)^{-0.4} f(u)\,du\right\}$$

In general,

$$_0D_x^{\alpha}f(x) = D^m\left\{_0D_x^{-v}f(x)\right\} = D^m\left\{\frac{1}{\Gamma(v)}\int_0^x (x-u)^{v-1} f(u)\,du\right\} \tag{9.23}$$

Substituting $v=m-\alpha$ and taking $\dfrac{1}{\Gamma(v)}=\dfrac{1}{\Gamma(m-\alpha)}$ outside the differential gives the following equation:

$$_0D_x^{\alpha}f(x) = \frac{1}{\Gamma(m-\alpha)}D^m\left\{\int_0^x (x-u)^{m-\alpha-1} f(u)\,du\right\} \tag{9.24}$$

9.7.2 Caputo Fractional Derivative

Another type of fractional derivative is the Caputo fractional derivative of a causal function $f(x)$ (i.e., $f(x)=0$ for $x<0$), which is defined as follows:

$$_b^C D_x^{\alpha}f(x) = {_bI_x^{n-\alpha}}\frac{d^n}{dx^n}f(x) = {_bD_x^{-(n-\alpha)}}f^{(n)}(x) = \frac{1}{\Gamma(n-\alpha)}\int_b^x \frac{f^{(n)}(u)}{(x-u)^{\alpha+1-n}}\,du \tag{9.25}$$

where n is the nearest integer bigger than α. The operator $_b^C D_x^{\alpha}$ is sometimes denoted by $_bD_{*x}^{\alpha}$ and is called the *Caputo differential operator* of order α. From Equation 9.25 we have that the Caputo fractional derivative of a constant M is given by

$$_bD_{*x}^{\alpha}f(x) = {_b^C D_x^{\alpha}}f(x) = {_bD_x^{-(n-\alpha)}}f^{(n)}(x) = {_bD_x^{-(n-\alpha)}}\frac{d^n M}{dx^n} = 0$$

Recall that the RL fractional derivative of a constant M is given by

$$\frac{d^{\alpha}}{dx^{\alpha}}M = {_0D_x^{\alpha}}Mx = \frac{Mx^{-\alpha}}{\Gamma(1-\alpha)}$$

Thus, one of the advantages of the Caputo derivative is its ability to give the derivative of a constant as 0, as in ordinary derivative.

To compare the RL and Caputo fractional derivatives side by side, we reproduce their values respectively as follows. For the RL derivative, we have

$$D_x^\alpha f(x) = \begin{cases} \dfrac{1}{\Gamma(m-\alpha)} \dfrac{d^m}{dx^m}\left[\displaystyle\int_0^x \dfrac{f(u)\,du}{(x-u)^{\alpha+1-m}}\right] & m-1 < \alpha < m \\[4mm] \dfrac{d^m}{dx^m} f(x) & \alpha = m \end{cases}$$

Similarly, for the Caputo derivative we have

$$D_{*_x}^\alpha f(x) = \begin{cases} \dfrac{1}{\Gamma(m-\alpha)} \displaystyle\int_0^x \dfrac{f^{(m)}(u)\,du}{(x-u)^{\alpha+1-m}} & m-1 < \alpha < m \\[4mm] \dfrac{d^m}{dx^m} f(x) & \alpha = m \end{cases}$$

In summary, two of the limitations of the RL derivative are as follows:

1. The derivative of a constant is not 0, which might be a problem when using RL operators to write evolution equation. Specifically $_bD_x^\alpha A = \dfrac{A(x-b)^{-\alpha}}{\Gamma(1-\alpha)}$.

2. The Laplace transform of the RL derivative depends on the fractional derivatives at 0, which is usually an issue when solving initial-value problems:

$$L\left\{{}_0D_x^\alpha f\right\} = s^\alpha F(s) - \sum_{k=0}^{n-1} s^k \left[{}_0D_x^{\alpha-k-1} f\right]_{x=0} \quad n-1 \le \alpha < n$$

9.7.3 Grunwald–Letnikov Fractional Derivative

The GL definition is given by

$$_aD_t^\alpha f(t) = \lim_{h \to 0} \frac{1}{h^\alpha} \sum_{j=0}^{\left[\frac{t-\alpha}{h}\right]} (-1)^j \binom{\alpha}{j} f(t-jh)$$

where h is the time increment and $[\cdot]$ means the integer part of the argument. The GL definition caters for both fractional differentiation and fractional integration because positive values of α give fractional differentiation while negative values give fractional integration. Also, while RL and Caputo definitions provide a way to find analytical solutions, the GL definition provides numerical computation of results and thus is more suitable for numerical analysis.

9.8   FRACTIONAL DIFFERENTIAL EQUATIONS

Fractional differential equations (FDEs) involve fractional derivatives of the form d^α/dx^α, which are defined for $\alpha > 0$ where α is not necessarily an integer. They are generalizations of the ordinary differential equation (ODE) to a random (noninteger) order. They have attracted considerable interest due to their ability to model complex phenomena. The equation

$$D^\alpha f(x) = u(x, f(x)) \tag{9.26}$$

is called a *FDE of the RL type*. The initial conditions for this type of FDE are of the form

$$D^{\alpha-k} f(0) = b_k \quad k = 1, 2, \ldots, n-1$$
$$I^{n-\alpha} f(0) = b_n$$

Similarly, the equation

$$D_*^\alpha f(x) = u(x, f(x)) \tag{9.27}$$

is called a FDE of the Caputo type, and the initial conditions are of the form

$$D^k f(0) = b_k \quad k = 1, 2, \ldots, n-1$$

Consider the FDE based on the RL derivative:

$$_0D_t^\beta f = \frac{1}{\Gamma(1-\beta)} \frac{\partial}{\partial t} \int_0^t \frac{f(\tau)}{(t-\tau)^\beta} d\tau$$

Its Laplace transform is

$$L\left[_0D_t^\beta f\right] = s^\beta \tilde{f}(s) - \left[_0D_t^{-(1-)\beta} f\right](0)$$

The Laplace transform depends on the initial value of the fractional integral of f rather than the initial value of f that is typically given in physical applications. It is well-known that to solve classical and FDEs, we need to specify additional conditions in order to produce a unique solution. For RL FDEs, these additional conditions constitute certain fractional derivatives and/or integrals of the unknown solution at the initial point $x = 0$, which are functions of x. Unfortunately, these initial conditions are not physical and cannot be generally measured, and this is a major drawback of this type of fractional derivative. A solution of this problem is provided by using the Caputo derivative of the fractional derivative where these additional conditions are essentially the traditional conditions that are similar to those of classical differential equations that we are familiar with. Thus, the equation of choice in most cases is that based on the Caputo derivative that incorporates the initial values of the function and of its integer derivatives of lower order. A fractional derivative in the Caputo sense is defined as follows:

$$D_*^\alpha f(x) = I^{m-\alpha} D^m f(x) = \frac{1}{\Gamma(m-\alpha)} \int_0^x (x-u)^{m-\alpha-1} f^{(m)}(u)\, du \qquad (9.28)$$

where $m-1 < \alpha \le m$, $m \in N$, $x > 0$. The following properties apply

$$D_*^\alpha I^\alpha f(x) = f(x)$$

$$I^\alpha D_*^\alpha f(x) = f(x) - \sum_{k=0}^{m-1} f^{(k)}(0^+) \frac{x^k}{k!} \quad x > 0$$

Also, most of the applications of FDEs involve relaxation and oscillation models. We begin by reviewing the regular differential equations of these two models, which involve integer orders.

9.8.1   Relaxation Differential Equation of Integer Order

A relaxation differential equation is an initial-value differential equation of the form:

$$\frac{df(t)}{dt} = -cf(t) + u(t), \quad t > 0, \ f(0^+) = f_0 \qquad (9.29)$$

where c is a constant. The solution to this equation is

$$f(t) = f_0 e^{-ct} + \int_0^t u(t-\tau) e^{-c\tau} d\tau \qquad (9.30)$$

9.8.2   Oscillation Differential Equation of Integer Order

An oscillation differential equation is an initial-value differential equation of the form:

$$\frac{d^2 f(t)}{dt^2} = -f(t) + u(t), \quad t > 0, \ f(0^+) = f_0, \ f'(0^+) = f_1 \qquad (9.31)$$

It is called an oscillation differential equation because it has an oscillating sinusoidal solution as follows:

$$f(t) = f_0 \cos(t) + f_1 \sin(t) + \int_0^t u(t-\tau)\sin(\tau)\,d\tau \qquad (9.32)$$

9.8.3 Relaxation and Oscillation Fractional Differential Equations

The relaxation FDE has order $0 < \alpha \leq 1$ while the oscillation FDE has an order $1 < \alpha \leq 2$. However, unlike in the integer-order ODE there is no need to make a distinction between the two forms of FDE. Using the Caputo FDE, we can represent both forms as follows:

$$D_*^\alpha f(t) = D^\alpha \left\{ f(t) - \sum_{k=0}^{m-1} \frac{t^k}{k!} f^{(k)}(0) \right\} = -f(t) + u(t) \quad m-1 < \alpha \leq m \qquad (9.33)$$

where m is an integer. Since $I^\alpha D^\alpha f(t) = f(t)$, performing the I^α operation on the equation gives

$$f(t) = \sum_{k=0}^{m-1} \frac{t^k}{k!} f^{(k)}(0) - I^\alpha f(t) + I^\alpha u(t) \quad m-1 < \alpha \leq m \qquad (9.34)$$

The Laplace transform of the Mittag–Leffler function helps us to write the solution of FDEs in terms of this function. It can be shown that

$$L\left\{ t^{bk+\beta-1} E_{\alpha,\beta}^{(k)}\left(bt^\alpha\right) \right\} = \frac{k! s^{\alpha-\beta}}{\left(s^\alpha - b\right)^{k+1}}$$

$$L\left\{ t^{bk+\beta-1} E_{\alpha,\beta}^{(k)}\left(-bt^\alpha\right) \right\} = \frac{k! s^{\alpha-\beta}}{\left(s^\alpha + b\right)^{k+1}}$$

where

$$E_{\alpha,\beta}^{(k)}(x) = \frac{d^k}{dx^k} E_{\alpha,\beta}(x).$$

We can solve Equation 9.34 using the Laplace-transform method, which yields

$$F(s) = \sum_{k=0}^{m-1} \frac{1}{s^{k+1}} f^{(k)}(0) - \frac{1}{s^\alpha} F(s) + \frac{1}{s^\alpha} U(s)$$

This gives

$$F(s) = \sum_{k=0}^{m-1} \frac{s^{\alpha-k-1}}{s^\alpha + 1} f^{(k)}(0) + \frac{1}{s^\alpha + 1} U(s) = \sum_{k=0}^{m-1} \frac{1}{s^k} \left\{ \frac{s^{\alpha-1}}{s^\alpha + 1} \right\} f^{(k)}(0) + \frac{1}{s^\alpha + 1} U(s)$$

$$\qquad (9.35)$$

From the properties of the Laplace transform in Section 9.4, we have that

$$\frac{1}{s^k}\left\{\frac{s^{\alpha-1}}{s^\alpha+1}\right\} = L\left\{I^k E_\alpha\left(-t^\alpha\right)\right\}$$

Also, from Section 9.4 we have that

$$\frac{1}{s^\alpha+1} = L\left\{\frac{d}{dt}E_\alpha(-t^\alpha)\right\}$$

Thus, we have the solution to the FDE as

$$f(t) = \sum_{k=0}^{m-1} I^k E_\alpha\left(-t^\alpha\right)f^{(k)}(0) - u(t)*E'_\alpha\left(-t^\alpha\right) \tag{9.36}$$

9.9   APPLICATIONS OF FRACTIONAL CALCULUS

As stated earlier, fractional calculus has found application in several areas of science and engineering. The interest in these applications arises from the fact that the fractional operator takes into account the memory of the past evolution of the variable. This is in contrast with the standard integer-order differential operator that considers only the local time history. This feature poses some difficulties, both for the initialization and the numerical calculation of the fractional operator but, on the other hand, constitutes a major advantage over the standard differential scheme because it leads to a concise description of the system dynamics.

In this section, we give a brief description of these applications. Specifically, we consider the following fractional processes:

1. Fractional Brownian Motion
2. Fractional Random Walk
3. Fractional Diffusion
4. Fractional Gaussian Noise
5. Fractional Poisson Process

9.9.1   Fractional Brownian Motion

The fractional Brownian motion (FBM) $\{B_H(t),\ t \geq 0\}$ is a generalization of the Brownian motion. Although the main principles of FBM were introduced earlier by Kolmogorov, the name was introduced in Mandelbrot and van Ness (1968) who defined the process by the stochastic integral

$$B_H(t) = \frac{1}{\Gamma\left(H + \frac{1}{2}\right)} \left\{ \int_{-\infty}^{0} \left[(t-u)^{H-1/2} - (-u)^{H-1/2} \right] dB(u) + \int_{0}^{t} (t-u)^{H-1/2} \, dB(u) \right\}$$

$$(9.37)$$

for $t \geq 0$, $0 < H < 1$, and $B(t)$ is a Brownian motion. FBM is a centered Gaussian process with stationary but not independent increments; it has independent increments only when it is a standard Brownian motion. More formally, a random process $\{B_H(t), t \geq 0\}$ is an FBM if, for an index $0 < H < 1$ the following conditions hold:

1. $B_H(t)$ is continuous with probability 1 and $B_H(0) = 0$
2. For any $t \geq 0$ and $H > 0$, the increment $B_H(t+\theta) - B_H(t)$ is normally distributed with mean zero and variance θ^{2H}, which means that the CDF of $B_H(t+\theta) - B_H(t)$ is given by

$$P\left[\{ B_H(t+\theta) - B_H(t) \} \leq x \right] = \frac{1}{\sqrt{2\pi\theta^{2H}}} \int_{-\infty}^{x} \exp\left(-\frac{u^2}{2\theta^{2H}} \right) du \qquad (9.38)$$

The index H is called the *Hurst index* (or *Hurst parameter*), which is used to model the dependence property of the increments. If $H = 1/2$, then FBM becomes the normal Brownian motion. Thus, the main difference between regular Brownian motion and the FBM is that while the increments in the regular Brownian motion are independent, they are dependent in the FBM. This dependence means that if there is an increasing pattern in the previous "step," then the current step is likely to be increasing also. While FBM permits correlation between events, it retains the short tails of the normal distribution.

FBM satisfies the following condition:

$$B_H(0) = E\left[B_H(t) \right] = 0 \quad t \geq 0 \qquad (9.39)$$

The covariance function of FBM, $R_H(s, t)$, is given by

$$R_H(s, t) = E\left[B_H(s) B_H(t) \right] = \frac{1}{2}\left(s^{2H} + t^{2H} - |t-s|^{2H} \right) \quad s, t \geq 0 \qquad (9.40)$$

This result, which implies that the FBM is nonstationary, can be established as follows:

$$E\left[\{ B_H(s) - B_H(t) \}^2 \right] = E\left[\{ B_H(s-t) - B_H(0) \}^2 \right] = E\left[\{ B_H(s-t) \}^2 \right] = \sigma^2(s-t)^{2H}$$

Also,

$$E\left[\left\{B_H(s)-B_H(t)\right\}^2\right] = E\left[\left\{B_H(s)\right\}^2\right] + E\left[\left\{B_H(t)\right\}^2\right] - 2E\left[B_H(s)B_H(t)\right]$$
$$= \sigma^2(s)^{2H} + \sigma^2(t)^{2H} - 2E[B_H(s)B_H(t)]$$

Thus, $\sigma^2(s)^{2H} + \sigma^2(t)^{2H} - 2E\left[B_H(s)B_H(t)\right] = \sigma^2(s-t)^{2H}$, which gives

$$E\left[B_H(s)B_H(t)\right] = \frac{\sigma^2}{2}\left\{(s)^{2H} + (t)^{2H} - (s-t)^{2H}\right\}$$

where $\sigma=1$ for the standard Brownian motion. The value of H determines the kind of process the FBM is. Generally, three types of processes can be captured by the model:

- If $H>1/2$, the increments of the process are positively correlated, and the process is said to be *persistent* or have a *long memory* (or *strong aftereffects*). In this case, consecutive increments tend to have the same sign; that is, a random step in one direction is preferentially followed by another step in the same direction.
- If $0<H<1/2$, the increments are negatively correlated. In this case, the process is said to be *anti-persistent*, which means that consecutive increments are more likely to have opposite signs; that is, a random step in one direction is preferentially followed by a reversal of direction.
- If $H=1/2$, the process is a regular Brownian motion in which steps are statistically independent of one another. Thus, successive Brownian motion increments are as likely to have the same sign as the opposite, and thus there is no correlation.

Another property of FBM is self-similarity. As discussed in Chapter 8, a self-similar process is a stochastic process that is invariant in distribution under suitable scaling of time and space. Specifically, the stochastic process $\{X(t), t\geq 0\}$ is said to be self-similar if for any $a>0$ there exists some $b>0$ such that $\{X(at), t\geq 0\}$ has the same distribution as $\{bX(t), t\geq 0\}$. In the case of the FBM with a Hurst index H, a and b are related by $b=a^H$. Thus, FBM is a self-similar stochastic process in the sense that $\{B_H(at), t\geq 0\}$ has the same distribution as $\{a^H B_H(t), t\geq 0\}$ for all $a>0$.

Also, FBM exhibits long-range dependence when $H>1/2$, which means that if we define the autocorrelation function $R_H(n)=E[B_H(t_0)B_H(t_n)]$, then

$$\sum_{n=1}^{\infty} R_H(n) = \infty$$

When $H<1/2$, $R_H(n)$ tends to decay exponentially; that is, $R_H(n)=\beta^n$ as n tends to infinity, where $\beta<1$, which means that the process exhibits short-range dependence.

FBM has been used in Norros (1995) and Mikosch et al. (2002) to model aggregated connectionless traffic. It has also been used in Rogers (1997), Dasgupta (1998), and Sottinen (2001) to model financial markets. The motivation for using FBM to model financial markets is that earlier empirical studies indicated that the so-called logarithmic returns $r_n = \log\{S(t_n)/S(t_{n-1})\}$, where $S(t)$ is the observed price of a given stock at time t, is normal, which means that $H = 0.5$. However, new empirical studies indicate that r_n tends to have a strong aftereffect with $H > 0.5$. Thus, advocates of FBM claim that it is a better model than traditional Brownian motion by using the Hurst index to capture dependency.

9.9.2 Multifractional Brownian Motion

One of the disadvantages of FBM is that because the Hurst index, which governs the pointwise regularity of a system, is constant, it can only be used to model phenomena that have the same irregularities globally; that is, phenomena that have monofractal structure. Thus, when a system has intricate structures with variations in irregularities, FBM becomes inadequate and a model that permits the Hurst index to vary as a function of time or position becomes necessary. The multifractional Brownian motion (MBM) is used in this case. In MBM, the Hurst index is replaced by the *Holder function H(t)* that describes the local variations in the system where $0 < H(t) < 1$. Thus, MBM generalizes FBM by replacing the Hurst index with the Holder function. This generalization was introduced independently by Peltier and Vehel (1995) and Benassi et al. (1997). It can be defined as follows:

$$B_{H(t)}(t) = \frac{1}{\Gamma\left(H(t)+\frac{1}{2}\right)}\left\{\int_{-\infty}^{0}\left[(t-u)^{H(t)-1/2}-(-u)^{H(t)-1/2}\right]dB(u) + \int_{0}^{t}(t-u)^{H(t)-1/2}dB(u)\right\}$$

$$(9.41)$$

where $B(t)$ is a Brownian motion.

The fact that the local regularity of MBM may be tuned via a functional parameter has made it a useful model in various areas such as finance, biomedicine, geophysics, and image analysis. For example, when innovations of the time series of the rates of return in the stock market are independent, the time series can be modeled by a Brownian motion. If the series has long memory, then it is better modeled by an FBM. However, because stock prices experience some "quiet" periods that correspond to a higher Holder exponent and some "erratic" periods that correspond to a low Holder exponent, such financial time series are better modeled by a MBM.

9.9.3 Fractional Random Walk

Fractional random walk is a semi-random walk process that is a generalization of the random walk. The difference between the regular random walk and the fractional random walk lies in the fact that the latter takes the correlation between walking steps into account. Thus, fractional random walk has dependent movements and is used to model financial time series because it allows long memory found in stock returns to be captured.

In a standard random walk, the walker takes the ith step of length $X_i = 1$ with probability $\frac{1}{2}$ and a length $X_i = -1$ with probability $\frac{1}{2}$. The location of the walker after the nth step is given by

$$Y_n = \begin{cases} \sum_{i=1}^{n} X_i & n > 0 \\ 0 & n = 0 \end{cases}$$

Using the method used in Roman and Porto (2008), we consider the behavior of the associated fractional random walk obtained by applying the fractional operator $D_1^{(\alpha)}$ to Y_n; that is, $D_1^{(\alpha)} Y_n = Y_n^{(\alpha)}$. Let $\Delta Y_n = Y_n - Y_{n-1}$ be the increment of the standard random walk. This increment is a local function of the time step n; that is, it is independent of the previous values of X_i, $i < n$. Equivalently, let the increments in a fractional random walk be defined by

$$\Delta Y_n^{(\alpha)} = Y_n^{(\alpha)} - Y_{n-1}^{(\alpha)} \tag{9.42}$$

Consider the GL fractional derivative operator of order α, $-1 < \alpha < 1$, $D_\tau^{(\alpha)}$, which is defined as

$$D_\tau^{(\alpha)} Y_n = \frac{1}{\tau^\alpha} \sum_{i=0}^{n} (-1)^i \binom{\alpha}{i} Y_{n-i} \tag{9.43}$$

where the binomial coefficient

$$\binom{\alpha}{i} = \frac{\Gamma(\alpha+1)}{\Gamma(i+1)\Gamma(\alpha+1-i)} = \frac{\alpha(\alpha-1)(\alpha-2)\cdots(i-\alpha-1)}{i!}$$
$$= (-1)^i \frac{(0-\alpha)(1-\alpha)(2-\alpha)\cdots(\alpha-i+1)}{i!} \tag{9.44}$$

Using the relation

$$\prod_{k=0}^{i-1}(k+z) = \frac{\Gamma(i+z)}{\Gamma(z)}$$

in Equation 9.44 with $z = -\alpha$ and defining the coefficient

$$C_i^{(\alpha)} = (-1)^i \binom{\alpha}{i}$$

we obtain the following for the coefficient in Equation 9.44:

$$C_i^{(\alpha)} = \frac{\Gamma(i-\alpha)}{\Gamma(-\alpha)i!} = -\frac{\alpha}{\Gamma(1-\alpha)}\frac{\Gamma(i-\alpha)}{\Gamma(i+1)}$$

where we have used the relations $\Gamma(1+z)=z\Gamma(z)$ and $i!=\Gamma(i+1)$. Thus, we obtain

$$D_\tau^{(\alpha)}Y_n \equiv Y_n^{(\alpha)} = \frac{1}{\tau^\alpha}\sum_{i=0}^{n}C_i^{(\alpha)}Y_{n-i} \tag{9.45}$$

Applying this result to Equation 9.43 we obtain

$$\begin{aligned}
\Delta Y_n^{(\alpha)} &= \frac{1}{\tau^\alpha}\sum_{i=0}^{n}C_i^{(\alpha)}Y_{n-i} - \sum_{i=0}^{n-1}C_i^{(\alpha)}Y_{n-1-i} \\
&= C_n^{(\alpha)}Y_0 + \sum_{i=0}^{n-1}C_i^{(\alpha)}\left\{Y_{n-i}-Y_{n-1-i}\right\} \\
&= \sum_{i=0}^{n-1}C_i^{(\alpha)}\left\{Y_{n-i}-Y_{n-1-i}\right\}
\end{aligned} \tag{9.46}$$

since $Y_0=0$. The result indicates that for fractional α, the increments of the fractional random walk, $\Delta Y_n^{(\alpha)}$, require the knowledge of the whole history of the walk $\{X_i\}$ with $1 \le i \le n$, thus suggesting that FRWs may possess long-range autocorrelations.

9.9.4 Fractional (or Anomalous) Diffusion

Diffusion is macroscopically associated with a gradient of concentration. As we discussed in Chapter 7, it can be considered as the movement of molecules from a higher concentration region to a lower concentration region.

The basic property of the normal diffusion is that the second-order moment of the space coordinate of the particle $E[X^2]=\int x^2 f(x,t)\,dx$ is proportional to time, t, where $f(x,t)$ is the probability density function (PDF) of finding the particle at x at time t; that is,

$$E[X^2] \propto t^\alpha,$$

where $\alpha=1$. A diffusion process in which $\alpha \neq 1$ is called *anomalous diffusion*. Anomalous diffusion is also referred to as *fractional diffusion*. The one-dimensional normal diffusion equation and its corresponding solution are given by

$$\frac{\partial u(x,t)}{\partial t} = D\frac{\partial^2 u(x,t)}{\partial x^2} \tag{9.47}$$

$$u(x, t) = \frac{1}{\sqrt{4\pi Dt}} \exp\left\{-\frac{x^2}{4Dt}\right\} \tag{9.48}$$

If either one or both of the two derivatives are replaced by fractional-order derivatives, the resulting equation is called a fractional diffusion equation. We call the two types of fractional diffusion equation *time-fractional diffusion equation* and *space-fractional diffusion equation*. The time-fractional diffusion equation implies that the process has memory, while the space-fractional diffusion equation describes processes that are nonlocal. The *space-time fractional equation* considers both processes that have memory as well as being nonlocal.

Anomalous diffusion is classified through the scaling index α. When $\alpha = 1$, we obtain normal diffusion. When $\alpha > 1$, we obtain *superdiffusion* that includes the case when $\alpha = 2$, which is called *ballistic diffusion*. Superdiffusion is also called *enhanced diffusion*. When $0 < \alpha < 1$, we obtain *subdiffusion*, which is also called *suppressed diffusion*. Thus, an anomalous diffusion is a diffusion that is characterized by a mean square displacement that is proportional to a power law in time. This type of diffusion is found in many environments including diffusion in porous media, disordered solids, biological media, atmospheric turbulence, transport in polymers, and Levy flights. Fractional diffusion has been the subject of many reviews such as Haus and Kehr (1987), Bouchaud and Georges (1990), Havlin and Ben-Avraham (2002), Zaslavsky (2002), and Metzler and Klafter (2000, 2004), and other publications such as Vainstein et al. (2008).

In subdiffusion, the travel times of particles are much longer than those expected from classic diffusion because particles tend to halt between steps. Thus, particles diffuse slower than in regular diffusion. On the other hand, in superdiffusion, the particles spread faster than in regular diffusion. Superdiffusion occurs in biological systems while subdiffusion can be found in transport processes.

Fractional diffusion has been successfully modeled by the time-fractional diffusion equation that is obtained by replacing the integer order time derivative in the diffusion equation with a derivative of non-integer order, which gives

$$_0D_x^\alpha u(x, t) = \frac{\partial^\alpha u(x, t)}{\partial t^\alpha} = C_\alpha \frac{\partial^2 u(x, t)}{\partial x^2} \tag{9.49}$$

where $_0D_x^\alpha$ is the fractional RL derivative operator of order α.

Fractional diffusion equation can be derived from the continuous-time random walk (CTRW). Recall from Chapter 3 that CTRW is a random walk that permits intervals between successive walks to be independent and identically distributed. Thus, the walker starts at the point 0 at time $T_0 = 0$ and waits until time T_1 when he makes a jump of size x_1, which is not necessarily positive. The walker then waits until time T_2 when he makes another jump of size x_2, and so on. The jump sizes x_i are also assumed to be independent and identically distributed. Thus, we assume that the times T_1, T_2, ... are the instants when the walker makes jumps. The intervals $\tau_i = T_i - T_{i-1}$, $i = 1, 2, \ldots$, are called the *waiting times* (or *pausing times*) and are assumed to be independent and identically distributed.

Let T denote the waiting time and let X denote the jump size. Similarly, let $f_X(x)$ denote the PDF of X, let $f_T(t)$ denote the PDF of T, and let $P(x, t)$ denote the probability that the position of the walker at time t is x, given that it was in position 0 at time $t=0$; that is,

$$P(x, t) = P\left[X(t) = x \mid X(0) = 0\right] \qquad (9.50)$$

We consider an uncoupled CTRW in which the waiting time and the jump size are independent so that the master equation is given by

$$P(x, t) = \delta(x)R(t) + \int_0^t \int_{-\infty}^{\infty} P(u, \tau) f_X(x-u) f_T(t-\tau) \, du \, d\tau \qquad (9.51)$$

where $\delta(x)$ is the Dirac delta function and $R(t) = P[T > t]$ is called the *survival probability*, which is the probability that the waiting time when the process is in a given state is greater than t. The master equation states that the probability that $X(t) = x$ is equal to the probability that the process was in state 0 up to time t, plus the probability that the process was at some state u at time τ, where $0 < \tau \le t$, and within the waiting time $t - \tau$, a jump of size $x - u$ took place. Note that

$$R(t) = \int_t^{\infty} f_T(v) \, dv = 1 - \int_0^t f_T(v) \, dv$$

$$f_T(t) = -\frac{dR(t)}{dt}$$

When the waiting times are exponentially distributed such that $f_T(t) = \lambda e^{-\lambda t}, t > 0$, the survival probability is $R(t) = e^{-\lambda t}$ and satisfies the following relaxation ODE:

$$\frac{d}{dt}R(t) = -\lambda R(t), \quad t > 0, \quad R(0^+) = 1 \qquad (9.52)$$

The simplest fractional generalization of Equation 9.52 that gives rise to anomalous relaxation and power-law tails in the waiting time PDF can be written as follows:

$$\frac{d^\alpha}{dt^\alpha}R(t) = -R(t), \quad t > 0, \quad 0 < \alpha < 1; \quad R(0^+) = 1 \qquad (9.53)$$

where d^α/dt^α is the Caputo fractional derivative; that is,

$$\frac{d^\alpha}{dt^\alpha}R(t) = \frac{1}{\Gamma(1-\alpha)}\frac{d}{dt}\int_0^t \frac{R(u)}{(t-u)^\alpha} du - \frac{t^{-\alpha}}{\Gamma(1-\alpha)}R(0^+) \qquad (9.54)$$

Let $\tilde{a}$ denote the Laplace transform of a. Thus, taking the Laplace transform of Equation 9.53 incorporating Equation 9.54 we obtain

$$s^{\alpha}\,\tilde{R}(s) - s^{\alpha-1}R(0^{+}) = -\tilde{R}(s) \Rightarrow \tilde{R}(s) = \frac{s^{\alpha-1}}{1+s^{\alpha}} \qquad (9.55)$$

From Section 9.4, we have that the inverse Laplace transform of $\tilde{R}(s)$ is given by

$$R(t) = E_{\alpha}(-t^{\alpha}) \qquad (9.56)$$

Thus, the corresponding PDF of the waiting time is

$$f_{T}(t) = -\frac{dR(t)}{dt} = -\frac{dE_{\alpha}(-t^{\alpha})}{dt} = t^{\alpha-1}E_{\alpha,\alpha}(-t^{\alpha}) \qquad (9.57)$$

Let $M_{Y}(s)$ denote the Laplace transform (or s-transform) of the PDF $f_{Y}(y)$. Thus, from Equation 9.57 we have that

$$M_{T}(s) = \frac{1}{s^{\alpha}+1}, \quad 0 < \alpha \le 1 \qquad (9.58)$$

The asymptotic behavior of the PDF of the waiting time is as follows:

$$f_{T}(t) \sim \begin{cases} t^{\alpha-1} & t \to 0 \\ \dfrac{1}{t^{\alpha+1}} & t \to \infty \end{cases}$$

From this we observe that for $0 < \alpha < 1$ at large t, the function $f_{T}(t)$ does not decay exponentially anymore; instead it decays according to a power law. This means that because of the power-law asymptotic behavior of the process, it is no longer Markovian but of the long-memory type. Thus, the kind of diffusion that is associated with the CTRW depends on the distribution of the step increments. If the increments are small, we obtain normal diffusion. In this case, for jump sizes (or displacements) with finite variance σ^2 and waiting times with finite mean τ such we have that

$$\frac{\partial P(x, t)}{\partial t} = D_{0}\frac{\partial^{2}P(x, t)}{\partial x^{2}}$$

with a diffusion coefficient $D_{0} = \sigma^2/\tau$. Assume that both $f_{X}(x)$ and $f_{T}(t)$ exhibit algebraic tail such that

$$f_{X}(x) \sim |x|^{-(1+\beta)} \quad \text{and} \quad f_{T}(t) \sim t^{-(1+\alpha)},$$

for which σ^2 and τ are infinite. In this case, we can derive a space–time-fractional diffusion equation for the dynamics of $P(x, t)$:

$$\frac{\partial^{\alpha}P(x, t)}{\partial t^{\alpha}} = D_{\alpha,\beta}\frac{\partial^{\beta}P(x, t)}{\partial|x|^{\beta}}$$

where the constant $D_{\alpha\beta}$ is a generalized diffusion coefficient. Thus, the space-time fractional diffusion equation is obtained by replacing the first-order time derivative and second-order space derivative in the standard diffusion equation by a fractional derivative of order α and β, respectively.

As stated earlier, if we limit the power-law distribution to the waiting time, we obtain the time-fractional diffusion equation of the form:

$$\frac{\partial^\alpha P(x,t)}{\partial t^\alpha} = D_\alpha \frac{\partial^2 P(x,t)}{\partial x^2}$$

When $\alpha=1$, we obtain the classical diffusion equation. Similarly, if only the jump size (or displacement) has power-law distribution, we obtain the space-fractional diffusion equation of the form:

$$\frac{\partial P(x,t)}{\partial t} = D_\beta \frac{\partial^\beta P(x,t)}{\partial x^\beta}$$

In this case, when $\beta=2$, we obtain the classical diffusion equation. As stated earlier, the Caputo fractional derivative is generally used to solve FDEs because it allows traditional initial and boundary conditions to be included in the formulation in a standard way, whereas models based on other fractional derivatives may require the values of the fractional derivative terms at the initial time.

Time-Fractional Case: Consider the time-fractional diffusion equation

$$\frac{\partial^\alpha}{\partial t^\alpha} P(x,t) = D_\alpha \frac{\partial^2}{\partial x^2} P(x,t)$$
$$P(x,0) = \delta(x) \tag{9.59}$$
$$\lim_{x \to \pm\infty} P(x,t) = 0$$

Let $\hat{b}(w)$ denote the characteristic function of $b(x)$. That is,

$$\hat{b}(w) = \int_{-\infty}^{\infty} b(x) e^{jwx} dx \quad \text{and} \quad b(x) = \frac{1}{2\pi} \int_{-\infty}^{\infty} \hat{b}(w) e^{-jwx} dw$$

Recall that for a function $g(t)$, the Laplace transform of the Caputo derivative is

$$L\left\{ \frac{d^\alpha g(t)}{dt^\alpha} \right\} = s^\alpha \tilde{g}(s) - s^{\alpha-1} g(0^+)$$

Thus, applying the Laplace transform with respect to the time variable t and the characteristic function with respect to the space variable x in Equation 9.59 we obtain

$$s^\alpha \hat{\tilde{P}}(w, s) - s^{\alpha-1} = -D_\alpha w^2 \hat{\tilde{P}}(w, s) \Rightarrow \hat{\tilde{P}}(w, s) = \frac{s^{\alpha-1}}{s^\alpha + D_\alpha w^2}$$

Inverting the Laplace transform we obtain

$$\hat{P}(w, t) = L^{-1} \left\{ \frac{s^{\alpha-1}}{s^\alpha + D_\alpha w^2} \right\} = E_\alpha\left(-D_\alpha w^2 t^\alpha\right)$$

where $E_\alpha(\cdot)$ is the Mittag–Leffler function. Finally, inverting the characteristic function we obtain the probability $P(x, t)$ of having traversed a distance x at time t as follows:

$$P(x, t) = CF^{-1}\left\{ E_\alpha\left(-D_\alpha w^2 t^\alpha\right) \right\} = \frac{1}{2\pi} \int_{-\infty}^{\infty} e^{-jwx} E_\alpha\left(-D_\alpha w^2 t^\alpha\right) dw$$

Space-Fractional Case: We discuss the space-fractional diffusion equation through the Levy flights, which are random walk processes that have infinite second-order moments $E[X^2]$ and PDFs with heavy tails:

$$f_X(x) \sim \frac{1}{|x|^{1+\alpha}}, \quad 0 < \alpha < 2, \quad x \to \infty$$

Since the moments are infinite, we consider such processes as anomalous diffusion. Consider a CTRW with exponentially distributed waiting time, but a Levy step length whose asymptotic PDF has a power law:

$$f_X(x) = c \, | \, x \, |^{-1-\alpha}, \quad 1 < \alpha < 2$$

where c is a normalizing constant. Let $\Phi_x(w)$ denote the characteristic function of the PDF $f_X(x)$. Then the characteristic function of the preceding equation is

$$\Phi_X(w) = \exp\left(-c\,|w|^\alpha\right) \sim 1 - c\,|w|^\alpha \tag{9.60}$$

Applying the Laplace transform-characteristic function to Equation 9.51 we obtain

$$\hat{\tilde{P}}(w, s) = \frac{1 - M_T(s)}{s} + \hat{\tilde{P}}(w, s) M_T(s) \Phi_X(w)$$

which gives

$$\hat{\tilde{P}}(w, s) = \frac{1 - M_T(s)}{s} \frac{1}{1 - M_T(s)\Phi_X(w)}$$

When $f_T(t) = \lambda e^{-\lambda t}$ and we use the result in Equation 9.60, we obtain

$$\hat{\tilde{P}}(w, s) = \frac{1}{s + \lambda - \lambda \exp\left(-c|w|^{\alpha}\right)} = \frac{1}{s + \lambda \left[1 - \exp\left(-c|w|^{\alpha}\right)\right]}$$

Taking the inverse Laplace transform, we obtain

$$\hat{P}(w, t) = \exp\left\{-\lambda t\left[1 - \exp\left(-c|w|^{\alpha}\right)\right]\right\} \sim 1 - \lambda t\left[1 - \exp\left(-c|w|^{\alpha}\right)\right]$$

$$\sim 1 - \lambda t\left[1 - \left\{1 - c|w|^{\alpha}\right\}\right] = 1 - \lambda c t|w|^{\alpha}$$

Finally, taking the inverse characteristic function and recognizing that the equation is similar to Equation 9.60, we obtain the asymptotic long-time behavior as

$$P(x, t) \sim \frac{\lambda c t}{|x|^{1+\alpha}}, \quad 1 < \alpha < 2 \tag{9.61}$$

The solution of the space–time-fractional diffusion equation is more involved than that of either the time-fraction or space-fractional version. But the solution principles are the same.

9.9.5   Fractional Gaussian Noise

Fractional Gaussian noise (FGN) describes a time series that is somewhat random, but less random than white Gaussian noise. In an FGN process, there is long-range dependence, or long-time memory; that is, there is a significant correlation between observations of signals separated by large time spans. This means that what is happening at one instance has an effect on what happens later—even much later. Whereas the random walk is the discrete-time analog of the Brownian motion, the FGN is the discrete-time analog of the FBM.

More formally, FGN is the process $\{W_H(t), t > 0\}$ that is obtained from the FBM increments as follows:

$$W_H(t) = B_H(t+1) - B_H(t) \tag{9.62}$$

It is a stationary Gaussian process with zero mean and covariance given by

$$\rho(k) = E\left[W_H(t)W_H(t+k)\right] = \frac{1}{2}\left[|k+1|^{2H} - 2|k|^{2H} + |k-1|^{2H}\right], \quad k > 0 \tag{9.63}$$

When $H = 1/2$, all covariances are zero, except for $k = 0$, and $\{W_H(t), t > 0\}$ represents white Gaussian noise. By writing down the Taylor series expansion at the origin of the function $u(x)=(1-x)^{2H}-2+(1+x)^{2H}$ and noting that $\rho(k)=\dfrac{1}{2}k^{2H}u(1/k)$ for $k \geq 1$, it can be seen from Equation 9.63 that

$$\lim_{k \to \infty} \rho(k) \sim H(2H-1)k^{2H-2} \tag{9.64}$$

Thus, for $1/2<H<1$ we have that $\sum_{k=-\infty}^{\infty}\rho(k)=\infty$ because the correlations are not summable, which implies that the process has long-range dependence (or long memory). However, when $0<H<1/2$ we have that $\sum_{k=-\infty}^{\infty}\rho(k)=0$ because the correlations are summable, which means that the process has short-range dependence (or short memory).

9.9.6 Fractional Poisson Process

From Chapter 2, we know that one of the ways to define the Poisson process is that it is a counting process $\{N(t),\ t \geq 0\}$ in which the number of events in any interval of length t has a Poisson distribution with mean λt. Thus, its PMF is given by

$$p_{N(t)}(n,\ t)=P[N(t)=n]=\frac{(\lambda t)^{n}}{n!}e^{-\lambda t} \quad n=0,1,2,\ldots;\quad t \geq 0,\ \lambda>0$$

The parameter λ is essentially the mean number of arrivals per unit time or the arrival rate of the "customers." The time T between arrivals is a random variable that is usually called the waiting time whose PDF is given by

$$f_{T}(t)=\lambda e^{-\lambda t},\quad t \geq 0$$

The Poisson process is widely used in system modeling because it has no memory. The survival probability, $R(t)$, which is the probability that T is greater than t, is given by

$$R(t)=P[T>t]=e^{-\lambda t},\quad t \geq 0$$

The sum N_{k} of k independent and identically distributed exponential waiting times is the so-called Erlang random variable of order k whose PDF is given by

$$f_{N_{k}}(t)=\frac{\lambda^{k}t^{k-1}}{(k-1)!}e^{-\lambda t},\quad t \geq 0,\ k=1,2,\ldots$$

The basic idea of the fractional Poisson process $\{N_{\mu}(t),\ t \geq 0,\ \mathrm{Re}(\mu)>0\}$ is to make the standard Poisson model more flexible by permitting non-exponential, heavy-tailed distributions of inter-arrival times and different scaling properties. This is motivated by experimental data with long memory that is observed in complex systems.

The PMF of the fractional Poisson distribution is given by

$$p_{N_\mu(t)}(n, t) = \frac{(\lambda t^\mu)^n}{n!} \sum_{k=0}^{\infty} \frac{(k+n)!}{k!} \frac{(-\lambda t^\mu)^k}{\Gamma(\mu[k+n]+1)}, \quad 0 < \mu \leq 1 \qquad (9.65)$$

Observe that when $\mu = 1$, we obtain the Poisson PMF. The mean of $N_\mu(t)$ can easily be shown to be

$$E[N_\mu(t)] = \sum_{n=0}^{\infty} n p_{N_\mu(t)}(n,t) = \frac{\lambda t^\mu}{\Gamma(\mu+1)} \qquad (9.66)$$

Also, the probability of no arrival by time t is given by

$$p_{N_\mu(t)}(0, t) = \sum_{k=0}^{\infty} \frac{(-\lambda t^\mu)^k}{\Gamma(\mu k+1)} = E_\mu(-\lambda t^\mu) \qquad (9.67)$$

This is the survival probability, and thus we have that

$$R_\mu(t) = P[T_\mu > t] = E_\mu(-\lambda t^\mu), \quad t \geq 0, \, 0 < \mu \leq 1 \qquad (9.68)$$

Therefore, the waiting has the PDF

$$f_{T_\mu}(t) = -\frac{d}{dt} R_\mu(t) = -\frac{d}{dt} E_\mu(-\lambda t^\mu) = \lambda t^{\mu-1} E_{\mu,\mu}(-\lambda t^\mu) \qquad (9.69)$$

This is the fractional generalization of the exponential PDF. The Laplace transform of $f_{T_\mu}(t)$ is

$$F_{T_\mu}(s) = \frac{\lambda}{s^\mu + \lambda}, \quad 0 < \mu \leq 1 \qquad (9.70)$$

The asymptotic behavior of the PDF of the waiting time is as follows:

$$f_{T_\mu}(t) \sim \begin{cases} \lambda t^{\mu-1} & t \to 0 \\ \dfrac{1}{\lambda t^{\mu+1}} & t \to \infty \end{cases} \qquad (9.71)$$

From this we observe that for $0 < \mu < 1$ at large t, the function $f_{T_\mu}(t)$ does not decay exponentially anymore; instead, it decays according to a power law. This means that because of the power-law asymptotic behavior of the process, it is no longer Markovian but of the long-memory type. Note also that the asymptotic behavior of the survival probability is as follows:

$$R_\mu(t) = E_\mu(-\lambda t^\mu) \sim \begin{cases} 1 - \dfrac{\lambda t^\mu}{\Gamma(1+\mu)} + O(t^{2\mu}) & t \to 0 \\[4mm] \dfrac{1}{\lambda t^\mu \Gamma(1-\mu)} + O(t^{-2\mu}) & t \to \infty \end{cases} \qquad (9.72)$$

9.10 SUMMARY

Fractional calculus is the branch of calculus that generalizes ordinary calculus by letting differentiation and integration be of any arbitrary real or complex order. It has become popular due to the fact that while many analytical models assume that there is no memory in the system under investigation, observed data suggests that many of these systems actually exhibit some form of memory. Fractional calculus has proved to be a natural tool for modeling memory-dependent phenomena. It has found application in such areas as fluid flow, dynamic processes in self-similar and porous structures, diffusive transport, electrical networks, chemical physics, and signal processing.

This chapter has discussed the basic principles of fractional calculus including three definitions of fractional integro-differentials, which are the RL definition, the Caputo definition, and the GL definition. Different applications of fractional calculus have also been discussed, which are the FBM, MBM, fractional random walk, fractional diffusion, FGN, and fractional Poisson process.

CHAPTER 10

PERCOLATION THEORY

10.1 INTRODUCTION

Percolation theory was introduced by Broadbent and Hammersley (1957) to model
the flow of fluids in a porous medium with randomly blocked channels. Specifically,
the theory developed from the formulation of a stochastic model to answer the following question: What is the probability that a large porous stone immersed in a
bucket of water will be wet at its center? It has the property of exhibiting a *phase
transition*, and the occurrence of this *critical phenomenon* is central to its appeal.
Percolation theory has formally developed into a branch of probability theory that
deals with properties of random (or disordered) media, such as porous rocks, turbulent
fluids, and plasmas. It can also be considered the study of how systems of discrete
objects are connected to each other. In fact, percolation theory can be regarded as the
study of the connectivity of networks.

In percolation theory, the topology is a d-dimensional structure with the most popularly used topology being regular grids or lattices. It is mathematically attractive
because it exhibits relations between probabilistic and topological properties of
graphs. Thus, it is similar to random graph theory and differs from the latter primarily
in the terminology used. Vertices in random graph theory are called *sites* in percolation theory, and edges in random graph theory are called *bonds* in percolation theory.
Similarly, subgraphs are called *clusters* in percolation theory. A site can be in one of
two states: *open* or *closed*. Similarly, a bond can be open or closed. A site or bond is

Elements of Random Walk and Diffusion Processes, First Edition. Oliver C. Ibe.
© 2013 John Wiley & Sons, Inc. Published 2013 by John Wiley & Sons, Inc.

defined to be open if it is selected; otherwise it is defined to be closed. When a cluster is obtained by selecting vertices or sites, we have *site percolation*; and when a cluster is obtained by selecting edges or bonds, we have *bond percolation*.

Because of the complexity of percolation theory, most of the work on it is done via simulation. Therefore, our goal in this chapter is to present the basic principles of the process including phase transition and the concept of clustering. A cluster is simply a group of connected sites. An important question that arises relates to the size distribution of clusters that can be found when the occupation probability of a site in a lattice is *p*. Cluster size distributions are important in many areas of applications of percolation theory such as flow of liquid in a porous medium, conductor/insulator transition in composite materials, polymer gelation, and vulcanization. Other applications areas include social models, forest fires, biological evolution, and the spread of diseases in a population.

Finally, because percolation theory has a lot in common with random graph theory, we begin the chapter with an introduction to random graph theory.

10.2 GRAPH THEORY REVISITED

In Chapter 3, we defined a graph as a set of points (or *vertices*) that are interconnected by a set of lines (or *edges*). For a graph G we denote the set of vertices by V and the set of edges by E and thus write $G = (V, E)$. The number of vertices in a graph is $|V|$ and the number of edges is $|E|$. The cardinality $|V|$ is called the *order* of the graph, and the cardinality $|E|$ is called the *size* of the graph. An edge is specified by the two vertices that it connects. These two vertices are the endpoints of the edge. Thus, an edge whose two endpoints are v_i and v_j is denoted by $e_{ij} = (v_i, v_j)$. If a vertex is an endpoint of an edge, the edge is sometimes said to be *incident* with the vertex.

In this section, we extend the discussion to include topics on graph theory that will be encountered in this chapter.

10.2.1 Complete Graphs

A *complete graph* K_n on n vertices is the simple graph that has all $\binom{n}{2} = n(n-1)/2$ possible edges. The *density* of a graph is the ratio of the number of edges in the graph to the number of edges in a complete graph with the same number of vertices. Thus, the density of the graph $G = (V, E)$ is $2|E|/|V|\{|V|-1\}$. Figure 10.1 illustrates examples of complete graphs.

A *clique* of a graph G is a single node or a complete subgraph of G. That is, a clique is a subgraph of G in which every vertex is a neighbor of all other vertices in the subgraph. Figure 10.2 shows examples of cliques.

10.2.2 Random Graphs

A random graph is a finite set of vertices where edges connect vertex pairs in a random manner. *Bernoulli random graphs* are the earliest random graphs, and a graph $G(N, p)$ is called a Bernoulli random graph if it has N vertices and a vertex is

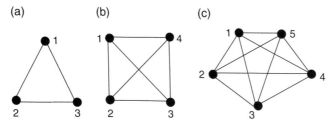

FIGURE 10.1   Examples of complete graphs. (**a**) K_3 complete graph; (**b**) K_4 complete graph; (**c**) K_5 complete graph.

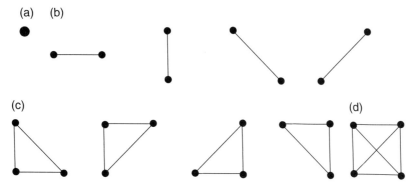

FIGURE 10.2   Examples of cliques. (**a**) One-node clique; (**b**) two-node cliques; (**c**) three-node cliques; (**d**) four-node clique.

connected to another vertex with probability p, where $0 \le p \le 1$. It is assumed that the connections occur independently. Thus, if we assume that the edges are independent and that self-loops are forbidden, the probability that the graph has m edges is given by

$$\binom{M}{m} p^m (1-p)^{M-m}$$

where $M = \binom{N}{2} = N(N-1)/2$ is the total possible number of edges.

Consider a vertex in the graph. It is connected with equal probability to each of the $N-1$ other vertices, which means that if K denotes the number of vertices that it is connected with, the probability mass function (PMF) of K is given by

$$p_K(k) = P[K = k] = \binom{N-1}{k} p^k (1-p)^{N-1-k} \qquad k = 0, 1, \ldots, N-1 \quad (10.1)$$

Thus, the expected value of K is $E[K] = (N-1)p$, which is the average degree of a vertex. Observe that the average degree is not bounded as $N \to \infty$, which is a limitation of the model. If we hold the mean $(N-1)p=\beta$ so that $p=\beta/(N-1)$, then the PMF becomes

$$p_K(k)=\binom{N-1}{k}\left[\frac{\beta}{N-1}\right]^k\left[1-\frac{\beta}{N-1}\right]^{N-1-k} \tag{10.2a}$$

$$\lim_{N\to\infty}p_K(k)=\frac{\beta^k}{k!}e^{-\beta} \tag{10.2b}$$

We recognize the last quantity as the Poisson distribution. A random graph in which each link forms independently with equal probability and the degree of each vertex has the Poisson distribution is called a *Poisson random graph*.

A component of a graph is a subset of vertices each of which is reachable from the others by some path through the graph. In the Bernoulli random graph, almost all vertices are disconnected at small values of p. As p increases, the number of clusters increases, where a cluster is a subgraph such that each vertex in the subgraph can be reached from any other vertex in the subgraph via a path that lies entirely within the subgraph and there is no edge between a vertex in the subgraph and a vertex outside of the subgraph. Generally, below some threshold probability $p=p_c$, the graph consists of only small disconnected clusters. Also, the average number of edges per vertex is less than one. However, above $p=p_c$, a *phase transition* occurs such that a single cluster or *giant component* emerges. Other clusters become vanishingly small in comparison to the giant component.

10.3 PERCOLATION ON A LATTICE

As discussed earlier, the space in percolation theory is usually represented as a lattice that is often infinite in extent. The points of intersection on the lattice are called *sites*, and the connections between sites (or the edges of the squares) are called *bonds*. As discussed earlier, in graph-theoretic language the sites are the nodes or vertices and the bonds are the edges of a graph.

From our earlier discussion, there are two common models of percolation: *bond percolation* and *site percolation*. Bond percolation is based on a grid of sites, where each site can be connected to each of four neighbors by a bond. Thus, the sites are fixed, and each bond is either porous (i.e., connected) or nonporous (i.e., not connected). A set of sites that are connected (directly or indirectly) by porous bonds is called a *cluster*. In the vocabulary of graphs, a cluster is a connected subgraph. Thus, in bond percolation, it is assumed that the sites are fixed and each site is randomly connected to its neighbor via bonds thereby forming a random graph.

On the other hand, site percolation is based on a random grid of sites and if two sites are adjacent, then they are considered connected. A set of connected sites is considered to be a cluster. Technically, the difference between the two types of percolation is that in bond percolation the sites are fixed and each bond is present with

(a) (b)

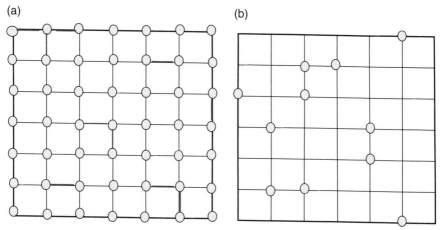

FIGURE 10.3   Examples of bond percolation and site percolation.

probability p and nonexistent with probability $1-p$; and in site percolation the sites are generated in a probabilistic manner such that each site is occupied with probability p and unoccupied with probability $1-p$. In site percolation, we refer to p as the *occupation probability*. Figure 10.3 illustrates the difference between bond percolation (Figure 10.3**a**) and site percolation (Figure 10.3**b**).

The main object of study in percolation is the cluster, which defines the connectivity properties of the sites on the lattice. For example, the rate of fluid flow in a percolation "network" is primarily determined by whether or not the porous sites form a path all the way through the material, which means that we need to know whether or not a set of sites (or bonds) contains a spanning cluster that includes two sites of interest.

10.3.1   Cluster Formation and Phase Transition

Each site (or bond) of a lattice is assumed to be occupied independently of its neighbors with a probability p that is also called the *concentration*. As stated earlier, a site and its neighbor belong to the same cluster if they are both occupied. A site is isolated if all its neighbors are empty. When the probability p is small, only small clusters will exist. Figure 10.4 shows an example of cluster formation in a site percolation system, where a cluster is a group of neighboring occupied sites with no empty sites between them.

With increasing values of p, the number of sites per cluster will increase. At a certain probability called the *critical probability* or *percolation threshold* denoted by p_c, a phase transition takes place and one *giant* (or *infinite*) *cluster* is formed that spans many of the sites (or bonds). Any site or bond that is not included in this giant cluster will be mainly an isolated site or a dangling bond. The giant cluster is called the *percolation cluster*, and the regimes $0 \le p < p_c$ and $p_c < p \le 1$ are called the regimes below and above the percolation threshold, respectively. Percolation is fundamentally

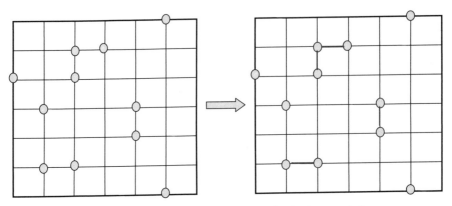

FIGURE 10.4 Example of cluster formation in site percolation.

FIGURE 10.5 Example of 1-cluster, 2-cluster, and 3-cluser in 1D lattice.

concerned with the analysis of clusters—their statistics and their properties. The exact determination of p_c is usually difficult and has been achieved for relatively few graph configurations. Much effort has gone into finding good upper and lower bounds for p_c.

The cluster size varies with p. Let a cluster with s sites be called an *s-cluster*. We first consider a one-dimensional lattice such as shown in Figure 10.5 where the black sites are unoccupied and the white sites are occupied. To have a 1-cluster, there must be an occupied site (with probability p) surrounded by two empty sites (with probability $(1-p)^2$, since each site behaves independently of other sites).

Thus, the probability of a 1-cluster is $p(1-p)^2$. If N is the total length of the chain and we ignore the end effect, the expected number of 1-clusters is $Np(1-p)^2$, and the expected number of 1-clusters per site is $p(1-p)^2$. If we define n_s as the number of s-clusters per site, it is obvious that

$$n_s = p^s(1-p)^2 \tag{10.3}$$

The expected number of occupied sites is Np, and any of these sites has the probability $n_s s$ to be a member of an s-cluster. Thus, we have that

$$\sum_{s=1}^{\infty} n_s s = \sum_{s=1}^{\infty} s(1-p)^2 p^s = (1-p)^2 \sum_{s=1}^{\infty} p \frac{d(p^s)}{dp}$$

$$= (1-p)^2 p \frac{d}{dp} \sum_{s=1}^{\infty} p^s = (1-p)^2 p \frac{d}{dp}\left(\frac{p}{1-p}\right)$$

$$= p \qquad\qquad\qquad p < p_c$$

Also, the probability that the cluster to which an occupied site belongs is an s-cluster is

$$w_s = \frac{n_s s}{p} = \frac{n_s s}{\sum\limits_{s=1}^{\infty} n_s s} \tag{10.4}$$

Thus, the mean cluster size is given by

$$
\begin{aligned}
E[S(p)] &= \sum_{s=1}^{\infty} s w_s = \sum_{s=1}^{\infty} \left\{ \frac{s^2 n_s}{\sum\limits_{s=1}^{\infty} n_s s} \right\} = \frac{1}{p} \sum_{s=1}^{\infty} s^2 n_s \\
&= \frac{1}{p}(1-p)^2 \sum_{s=1}^{\infty} s^2 p^s = \frac{1}{p}(1-p)^2 \left(p \frac{d}{dp} \right) \left(p \frac{d}{dp} \right) \left(\sum_{s=1}^{\infty} p^s \right) \\
&= \frac{1+p}{1-p}
\end{aligned}
\tag{10.5}
$$

Note that

$$\left(p\frac{d}{dp} \right)\left(p\frac{d}{dp} \right) = \left(p\frac{d}{dp} \right)^2 \neq p^2 \frac{d^2}{dp^2}$$

For a 1-dimensional lattice, $p_c = 1$, which means that the last result can be written as follows:

$$E[S(p)] = \frac{p_c + p}{p_c - p} \tag{10.6}$$

This shows that the mean cluster size becomes infinitely large as p approaches the critical probability, which is to be expected.

Another parameter of interest is the one-dimensional *correlation function* or *pair connectivity*, $g_1(r)$, which is the probability that a site that is a distance r from an occupied site is in the same cluster. For this to be true, all the $r-1$ intermediate sites must be occupied, which means that

$$g_1(r) = p^r \tag{10.7}$$

We now consider a two-dimensional square lattice. Figure 10.6 shows the different s-cluster configurations for $s=1$, 2, 3, and 4. The number of configurations of a 1-cluster is 1, the number of configurations of a 2-cluster is 2, the number of configurations of a 3-cluster is 6, and the number of configurations of a 4-cluster is 19. It

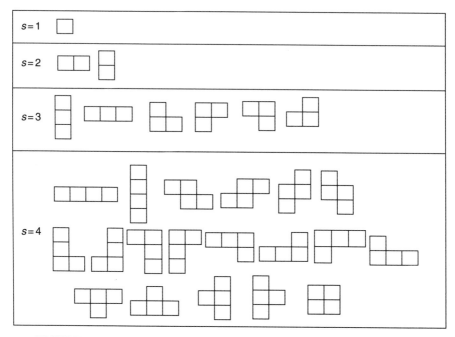

FIGURE 10.6   All cluster configurations on the square lattice for $s=k; k=1,\ldots,4$.

can be shown that the number of configurations of a 5-cluster is 63. Thus, the two-dimensional square lattice has a rather complicated cluster configuration because the number of configurations increases rapidly as s increases.

To compute the values of n_s, the number of s-clusters per site, we proceed as follows. On a square lattice, to have a 1-cluster, there must be an open site (with probability p) surrounded by four closed sites (with probability $(1-p)^4$). Thus, if L is the total number of sites in the lattice, the number of 1-clusters will be $Lp(1-p)^4$. Similarly, for 2-clusters, the number of such "left-right" clusters is $Lp^2(1-p)^6$, where the factor $(1-p)^6$ arises from the fact that the two occupied sites are surrounded by six unoccupied sites. This value is the same number of "up-down" clusters. The 3-cluster has two different types of configurations: in one type of configuration, the three occupied sites are surrounded by eight unoccupied sites; and in the other type of configuration, there are two occupied sites in either the horizontal or vertical direction and they are surrounded by seven unoccupied sites. Thus, we have that

$$n_1 = p(1-p)^4$$
$$n_2 = 2p^2(1-p)^6$$
$$n_3 = 2p^3(1-p)^8 + 4p^3(1-p)^7$$

The number of unoccupied neighbors of a cluster is called the *perimeter t* of the cluster. If the number of configurations with size s and perimeter t is denoted by g_{st}, then we obtain

$$n_s = \sum_t g_{st} p^s (1-p)^t$$

Unfortunately, the aforementioned equation is difficult in the general case since it involves the sum over all possible perimeters.

10.3.2 Percolation Probability and Critical Exponents

Another parameter of interest is the *percolation probability*, $\theta(p)$. We use the notation $u \leftrightarrow v$ to indicate that a path exists between sites u and v. The *open cluster* $C(x)$ of site x is defined as the set of occupied sites that are connected to x by a path. That is,

$$C(x) = \{u \in V : u \leftrightarrow x\}$$

If x is in no open path, then we let $C(x) = \{x\}$. We denote the number of sites in $C(x)$ by $|C(x)|$. As the occupation probability, p, is increased from 0, a typical cluster becomes larger (both in terms of the number of sites and geometric size). Because Z^2 (i.e., the plane square lattice) and the probability distribution are invariant to translation, statements about the cluster at an arbitrary site x are equivalent to statements about the cluster at the origin. Thus, the percolation probability, $\theta(p)$, is defined as the probability that the origin is part of an infinite cluster. That is, if we define $C(0)$ as a cluster that includes the origin, then

$$\theta(p) = P\left[|C(0)| = \infty\right] \tag{10.8}$$

Below $p = p_c$, $\theta(p)$ is always 0. As p is increased above p_c, the percolation probability becomes nonzero. This change in behavior is an example of a phase transition. When $p > p_c$, $\theta(p) \to 1$ and the system is said to be in the *supercritical* state so that clusters of infinite extent can be found. Similarly, when $p < p_c$, the system is said to be in the *subcritical* state where $\theta(p) = 0$. Thus, we can formally define p_c as follows:

$$p_c = \sup\{p : \theta(p) = 0\}$$

The function $\theta(p)$ is nondecreasing in p, and a conceptual graph is shown Figure 10.7. From the Figure we have that

$$\theta(p) = P\left[|C(0)| = \infty\right] = \begin{cases} = 0 & \text{if} \quad p < p_c \\ > 0 & \text{if} \quad p > p_c \end{cases} \tag{10.9}$$

Alternatively, we may write

$$\theta(p) \sim (p - p_c)^\beta, \qquad p > p_c \tag{10.10}$$

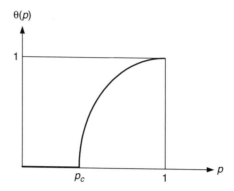

$\theta(p)$

FIGURE 10.7   The behavior of $\theta(p)$ around the critical probability, p_c.

for some positive β. This is known as a *power law* or *scaling law* and the exponent β is known as a *critical exponent*. This has the feature that it is entirely independent of the kind of lattice being studied or whether it is bond or site percolation. It only depends on the dimensionality of space (i.e., whether it has two dimensions or three dimensions). This is known as the *universality* feature, which is an important aspect of percolation theory. In the case of a two-dimensional percolation, $\beta = 5/36 \approx 0.139$, and in the case of three-dimensional percolation, $\beta = 0.41$.

In general, near the critical point p_c, many cluster properties vary as $|p - p_c|$ to a power. These powers that describe the properties of the percolating system at or near the threshold are the *critical exponents of percolation*. One of them deals with the size of the clusters. We first define the two-point *correlation function* $g(r)$ as the probability that if one point is in a cluster then another point at a distance r away is in the same cluster. This is the multidimensional analog of $g_1(r)$, which is defined in Equation 10.7. The function $g(r)$ typically has an exponential decay given by a *correlation length* $\xi(p)$; that is,

$$g(r) \sim e^{-r/\xi(p)}$$

At low values of p, the correlation length is small since clusters are typically of size one or two. As p increases, $\xi(p)$ increases until the threshold p_c when the spanning cluster dominates and is infinite in size, which means that the cluster size diverges. As the clusters get absorbed into the spanning cluster, the "typical" size of those left goes back down again. This means that cluster size first increases, then diverges at the threshold, and later decreases again. Thus, it is inversely proportional to a function of the deviation of p from the threshold p_c, and can be described in a mathematical form as follows:

$$\xi(p) \sim |p - p_c|^{-\nu} \tag{10.11}$$

where $v>0$ is another critical exponent. As with β, v it is universal (i.e., independent of details of the lattice) but depends on the dimension of the system. In two-dimensional percolation, $v=4/3=1.333$, and in three-dimensional percolation, $v=0.88$.

Another critical exponent is that associated with the average size of a finite cluster. The average cluster size is defined by $\chi(p)=E[|C(0)|:|C(0)|<\infty]$, where the average is taken over all configurations with the exception of those with an infinite cluster. This condition is imposed in order for $\chi(p)$ to remain finite when $p>p_c$. Although there is no infinite cluster at $p=p_c$, $\chi(p)$ diverges at this point because the cluster size distribution decays slowly. Thus, because the average cluster size diverges at $p=p_c$, we have that

$$\chi(p) \sim \left| p - p_c \right|^{-\gamma} \tag{10.12}$$

for some critical exponent $\gamma>0$.

The percolation threshold p_c plays a crucial role in the theory of percolation. The question on whether the interior of a lump of porous stone that is immersed in water will be wetted can now be answered from the point of view of percolation: When the site occupation probability p is larger than p_c the fluid may flow from one side of the rock to another. In the same way, when $p>p_c$, an epidemic or disease may spread over all individuals of a population, fire may destroy all trees in a model forest, and so on. Thus, the evaluation of the percolation threshold for many networks and lattices continues to appeal to scientists.

10.4   CONTINUUM PERCOLATION

Gilbert (1961) defined a model of continuum percolation in which points are no longer placed at the vertices of a lattice, but are randomly scattered in the plane according to a homogeneous Poisson point process of intensity $\lambda>0$. A point process is a stochastic process in which any one realization consists of a set of isolated points either in time or geographical space. There are two common models of continuum percolation, which are the *Boolean model* and the *random connection model*. Both models are based on points being generated according to a Poisson point process.

10.4.1   The Boolean Model

The Boolean model is driven by a Poisson point process X and each point of the process is the center of a ball of random radius where the radii associated with different points are independent and identically distributed, and the radii are independent of X. The region in the space that is covered by at least one ball is called an *occupied region*; otherwise it is defined as a *vacant region*. We denote by (X, r, λ) the Boolean model obtained from a Poisson process X of density λ and radius random variable r. For each point $x_i \in X$, we draw a sphere of radius r_i. Clusters of overlapping spheres will be obtained, as shown in Figure 10.8, and we are interested in the existence of an infinite cluster. Standard results show that

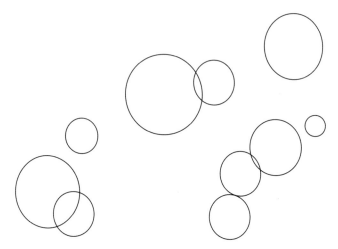

FIGURE 10.8 Example of the Boolean model.

there is a critical density λ_c such that a giant cluster occurs if $\lambda > \lambda_c$ and does not occur if $\lambda < \lambda_c$. Thus, when $\lambda > \lambda_c$, we say that the model percolates; that is, either there is a spanning cluster between two points of interest or there is a giant cluster that spans most of the points.

10.4.2 The Random Connection Model

As in Boolean models, the Poisson process is the first characteristic of the model and it assigns randomly points in the space. The second characteristic of the model is the *connection function*, which is a nonincreasing function from the positive real numbers onto [0, 1]. That is, a connection function $g : R^+ \to [0,1]$ connects two points $x_1, x_2 \in X$ with probability $g(|x_1 - x_2|)$, where $|\cdot|$ denotes the Euclidean distance. We denote the model by (X, g, λ). Thus, the construction of the model proceeds as follows. Given a connection function g, then for any two points x_1 and x_2 of the point process X, insert an edge between x_1 and x_2 with probability $g(|x_1 - x_2|)$, independently of all other pairs of point of X. Two points x and y of the model are said to be connected if there exists a finite sequence $(x = x_1, x_2, \ldots, x_n = y)$ such that the edge (x_i, x_{i+1}) exists for all $i = 1, \ldots, n-1$. As in the Boolean model, studies indicate that there exists a critical density λ_c such that when $\lambda > \lambda_c$, the random connection model percolates.

Note that the random connection model can be considered a special case of the Boolean model if spheres of radius r centered at each point overlap, which corresponds to the connection function

$$g(|x_1 - x_2|) = \begin{cases} 1 & \text{if } |x_1 - x_2| \leq 2r \\ 0 & \text{if } |x_1 - x_2| > 2r \end{cases} \tag{10.13}$$

10.5   BOOTSTRAP (OR *k*-CORE) PERCOLATION

Bootstrap percolation (BP) was proposed by Chalupa et al. (1979) to model dilute magnets in which, under some circumstances, a magnetic atom displays a magnetic moment only if it has enough magnetic neighbors. Each site is independently occupied with probability p, as in the regular model of percolation we have discussed so far. Then all occupied sites with less than a given number k of occupied neighbors are successively removed until either (a) a stable configuration of occupied sites is obtained such that each site has k or more occupied neighbors and no more sites can be removed or (b) all sites of the lattice are unoccupied. When $k=1$, isolated sites are removed; and when $k=2$, both isolated sites and dangling ends of clusters are removed. For sufficiently high k, no occupied sites remain for an infinite lattice. The case $k=0$ is by definition ordinary percolation on any lattice where all sites remain once they are occupied.

For $k=1$ or 2, the BP threshold $p_c(k)$ for any lattice is the same as the ordinary percolation threshold $p_c(0)$ because any infinite cluster is guaranteed to contain an infinite 2-cluster (and a 2-cluster is of course also a 1-cluster). Furthermore, when $k=1$, the critical behavior of all percolation functions (percolation probability, cluster size distribution, etc.) is the same as in ordinary percolation on all lattices because no cluster of two or more sites is culled or modified.

BP is also called the k-core percolation and is used to model those systems whose constituents require a minimum number of k connections to each other in order to participate in any clustering phenomenon. One method of constructing a k-core percolation system is as follow. Remove from a graph all sites of degree less than k. Some of the rest of the sites may remain with less than k neighbors. Then remove these sites, and so on, until no further removal is possible. The result, if it exists, is the k-core percolation system.

Given a particular lattice, the behavior of the BP probability can sometimes be deduced from simple geometrical arguments. In some cases, the sites that survive the removal process must be polygons or polyhedra. For example, when $m=3$ on the square lattice, surviving sites are rectangular. The primary interest in BP is the probability that at the end of the dynamics, that is, in the infinite time configuration, all the sites are occupied.

BP can be applied in the following situations:

- Some systems in solid state physics and chemistry can be presented as atoms occupying sites with probability p. An atom would participate in an ordered state like ferromagnetism only if a certain number of neighboring atoms were also present.
- In fluid flow in porous media, crack development and the advance of the fluid front are assumed strongly dependent on nearby pores and microstructure.

10.6   DIFFUSION PERCOLATION

Diffusion percolation (DP) deals with the dynamic growth of percolation clusters. The concept was developed by Adler and Aharony (1988) to model crack development when water flows through porous rocks. Both percolation and diffusion deal with fluids in a medium, but they differ in their emphasis. In diffusion, the emphasis is on the randomness of the fluid, while in percolation theory the emphasis is on the randomness of the medium. Specifically, percolation involves the spreading of a fluid in a random medium whereas in diffusion, the diffusing particle executes a random walk. While a position in the medium can be reached by diffusion, the spreading in percolation is confined to finite regions except when the percolation threshold is exceeded.

The scheme operates as follows. Initially, each site is occupied with probability p. Then in successive time steps an empty site with at least k occupied neighbors becomes occupied. The occupation process terminates when no more sites can be added. It is a model for the dynamic growth of percolation clusters and enhanced diffusion.

The problem of diffusion in a percolative lattice was coined by de Gennes (1976) as "the ant in the labyrinth." The ant (particle) may move from an occupied site to a nearest neighbor that is also occupied. At time $t=0$, the ant starts and looks to see if there is a nearest site that is occupied. If at least one such site exists, it chooses one at random and moves to that site. If there is no occupied nearest neighbor, the ant stays where it is. At time $t=1$, the process is repeated and so forth. After a time t the ant will have traveled a distance $R(t)$ from its initial position. We are interested in how the mean square displacement (MSD) $E[R^2(t)]$ depends on p.

The MSD will depend on the geometry of the lattice and on the value of p, the site occupation probability. When $p=0$, the ant has to stay where it is and cannot move so that $E[R^2(t)]=0$. When $p=1$, all sites are occupied and, are thus available to the ant, and the ant will perform a classical random walk with

$$E[R^2(t)] \sim t \tag{10.14}$$

We next consider intermediate values of p. Below the percolation threshold, $0<p<p_c$, all clusters are finite. This means that the ant can diffuse inside the cluster, but can never leave it. This clearly implies that

$$E[R^2(t)] \to \text{constant}, \qquad t \to \infty, \quad 0 < p < p_c \tag{10.15}$$

This constant will be a function of $|p_c - p|$. In the case of finite clusters, all sites of a cluster will be equally probable for long times. The probability that a site belongs to an s-cluster is sn_s. The size of such a cluster, thus, the distance over which the ant can travel, is R_s. Hence, the MSD of the random walk of the ant is

$$E[R^2(t)] = \sum_s sn_s R_s^2 \propto (p_c - p)^{\beta - 2\nu} \tag{10.16}$$

where β and ν are the critical exponents defined earlier. When the probability of occupation is larger than the percolation threshold ($p>p_c$), an infinite cluster exists and the ant may travel arbitrary far from its initial position. However, its "diffusive mobility" is limited for the following reasons:

1. There are finite clusters in the system that can act as traps for any ant that starts from a site inside such a cluster.

2. The infinite cluster may enclose many holes that will lead to many dead ends.

Consequently, near the percolation threshold the MSD will not grow linearly with time but will increase at a lower rate

$$E[R^2(t)] \sim t^{\alpha}, \qquad \alpha < 1, \quad p_c \le p < 1 \qquad (10.17)$$

This represents *sub-diffusion*, and the diffusion exponent α will be a function of the probability p and of the dimensionality d.

10.6.1   Bootstrap Percolation versus Diffusion Percolation

There is some similarity between DP and bootstrap percolation. Both of them are examples of *correlated percolation*. In BP, the sites are occupied at random with probability p and then each occupied site with less than k occupied neighbors is permanently changed to an empty site or "culled" from the lattice. This procedure continues until all occupied sites have at least k occupied neighbors or all sites are empty. The percolation threshold $p_{BP}(k)$ is the value of p at or above which an infinite cluster of occupied sites remains after the removal procedure. In DP, the sites are initially occupied with probability p. In successive time steps, empty sites with at least k occupied neighbors become occupied. In DP, the percolation threshold $p_{DP}(k)$ is the value of p at which an infinite cluster of occupied sites appears after the site-addition procedure.

10.7   FIRST-PASSAGE PERCOLATION

Consider a model where a certain node s in a graph represents an infection and that the edges of the graph may be open or blocked with given probabilities. In the Bernoulli percolation (or 0–1 percolation) that we considered earlier, we may be interested in such parameters as the probability that the infection can spread along open edges to reach a specific set T of target nodes. This means, for example, finding the probability that s and the T are in the same cluster. Suppose now that we still have the infected node s but that the infection takes a random time to spread along an edge to a neighboring node. In first-passage percolation, we are interested in parameters like the time taken for the infection to reach a set T of target nodes; that is, the first-passage time from s to T.

   To define the problem more formally, let $t(e)$ denote the passage time of edge e, which is the time it takes for fluid to flow along edge e. It is assumed that $t(e) \ge 0$, and that the $t(e)$ are independent and have a common cumulative distribution function (CDF) denoted by $F(t)$. For any path r that runs successively through edges e_1, $e_2, \dots, e_k$, we define the passage time of the path by

$$T(r) = \sum_{i=1}^{k} t(e_i)$$

One of the primary objects of interest in first-passage percolation is the first-passage time $T(x, y)$ between two sites x and y given by

$$T(x, y) = \min\{T(r) : r \text{ is a path from } x \text{ to } y\}$$

Another primary object of interest in first-passage percolation is the set of sites that can be reached from x by time t. This is given by

$$B(x, t) = \{y : T(x, y) \leq t\}$$

10.8 EXPLOSIVE PERCOLATION

Earlier in this chapter, we discussed the properties of a class of random graphs called the Bernoulli random graph, which is also called the Erdös-Rényi (ER) random graph. This graph can be obtained as follows. Starting with a collection of n nodes, where n is extremely large, connections between pairs of nodes are chosen at random one at a time and added to the graph. This evolving graph goes through a number of structural changes or phase transitions as it proceeds from the empty graph, which has no connections between pairs of nodes, to the complete graph in which all pairs of nodes are connected. One of these phase transitions in the graph's evolution is the emergence of the giant component. Specifically, assume that m edges have been added. Then for small values of m, typically $m < n/2$, many small subgraphs consisting of collections of nodes will be formed. As m increases, especially when $m > n/2$, a unique component that encompasses many of the nodes will be formed and all the other components will still be small. This single large component, which is called the giant component, slowly grows to cover the entire set of nodes as more connections are added.

The ER model and its phase transition are of fundamental importance for the following reasons. First, mathematically the properties of this phase transition are extremely well-understood, including the fine details of the behavior at and near the critical point, and the dynamics of the transition. For this reason, the model serves as the natural reference point when studying a wide range of phase transitions in many systems. Second, random graphs are the natural mathematical models for complex or disordered networks in the real world. For example, studying the emergence of large-scale connectedness as an edge-density parameter is varied corresponds to varying the proportion of links in an existing network that fail. Also, if the edges are used to model contacts that may spread a disease, then the emergence of a giant component corresponds to epidemic spread of a disease rather than localized outbreaks.

The evolution of the component structure has been studied in several random graph models. The interest in these models is not only to find out if and when the giant component appears but also how the giant component grows after it appears. As in the ER model, the emergence of the giant component in all of the models tends to

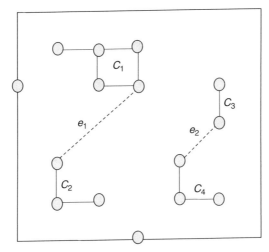

FIGURE 10.9   Illustration of the product rule.

mirror the phase transition in the ER random graph in several important ways. For example, the number of nodes in the giant component grows smoothly as more random connections are added.

A variant of the ER process that gained much attention over the last decade, mostly concerning the question on whether one can delay or accelerate the appearance of the giant component, is given by the so-called *Achlioptas process*. In this variant, we start with the empty graph on the vertex set, and in every step we are presented a pair of edges chosen uniformly at random from all pairs of vertices in the current graph. A fixed edge-selection rule then selects exactly one of them to be inserted into the graph while the other is put back into the pool of non-edges. A selection rule known as the *product rule* has been studied by Achlioptas et al. (2009) and has been shown to exhibit wildly different behavior. This rule operates as follows. Assume that there are four clusters: C_1 of size s_1, where the size is the number of nodes in the cluster; C_2 of size s_2; C_3 of size s_3; and C_4 of size s_4. Randomly choose a potential edge e_1 between C_1 and C_2, and a potential edge e_2 between C_3 and C_4. Then to delay the transition to the giant cluster choose e_1 if $s_1 \times s_2$ is smaller than $s_3 \times s_4$; otherwise choose e_2. Similarly, to enhance transition to the giant cluster, choose e_1 if $s_1 \times s_2$ is greater than $s_3 \times s_4$; otherwise choose e_2. Figure 10.9 illustrates the rule.

In Figure 10.9, we have that $s_1 = 5, s_2 = 3, s_3 = 2, s_4 = 3$. Thus, $s_1 \times s_2 = 15$ and $s_3 \times s_4 = 6$. Thus, to delay transition to the giant cluster we adopt the minimum product rule that chooses e_2, thereby merging C_3 and C_4 into one cluster of size 5. Figure 10.10 illustrates the onset of the transition to the giant component under the ER (or purely random) rule, the delayed (or minimum product) rule, and the enhanced (or maximum product) rule.

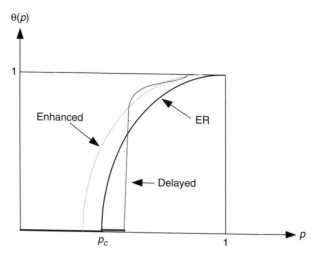

FIGURE 10.10   Onset of transition to giant component.

With the minimum product rule, the onset of the giant component is delayed but when it does begin, it emerges very suddenly, covering more than half of the nodes essentially immediately. This sudden growth in the size of the giant component is called explosive percolation. The minimum product rule was initially suggested by Bollobas (1984) as most likely to delay the phase transition. A series of papers has been devoted to understanding the phase transition in explosive percolation. However, most of the arguments are not rigorous but are only supported by simulations.

Other decision rules have been proposed for choosing the edge to add to the graph. They include the *sum rule* where we randomly select two edges that do not belong to the same cluster and connect them by comparing sums of the size of clusters as opposed to the products of the size of clusters. Another rule is the *Bohman–Frieze rule* proposed in Bohman and Frieze (2001), where an edge is chosen if it joins two isolated vertices, otherwise the other edge is chosen.

10.9   PERCOLATION IN COMPLEX NETWORKS

Complex networks are networks whose structure is irregular, complex, and dynamically evolving in time. They are encountered in a wide variety of disciplines and include technological networks, such as electrical power grids and the Internet; information networks, such as the World Wide Web; social networks, such as collaboration networks; and biological networks, such as neural networks and protein interaction networks.

Measurements have shown that these networks share two properties. The first fundamental network property is the fact that typical distances between vertices are small. This is the so-called *small-world phenomenon*. For example, in the Internet,

Internet Protocol (IP) packets traverse a few physical links. Thus, the graph of the Internet has evolved in such a way that typical distances are relatively small, even though the Internet is rather large. The second property of these networks is their *scale-free* nature, which means that they have power-law degree sequences; that is, the number of nodes with degree k falls off as an inverse power of k.

Although many quantities and measures of complex networks have been proposed and investigated in the last decades, three spectacular concepts play a key role in the recent study and development of complex networks theory:

- The average path length
- Clustering coefficient
- Degree distribution

10.9.1 Average Path Length

In a network, the distance d_{ij} between nodes i and j is defined as the number of edges along the shortest path connecting them. The *diameter D* of a network, therefore, is defined to be the maximal distance among all distances between any pair of nodes in the network. The *average path length L* of the network is defined as the mean distance between two nodes, averaged over all pairs of nodes. In a friendship network, for instance, L is the average number of friends existing in the shortest chain connecting two persons in the network. It has been discovered that the average path length of most complex networks is relatively small. This smallness property is called the *small-world effect*, and these networks are called *small-world networks*.

10.9.2 Clustering Coefficient

In a friendship network, it is quite possible that Miss X's friend's friend is also a direct friend of Miss X; or, to put it in another way, two of Miss X's friends are quite possibly friends of each other. This property refers to the *clustering* of the network. This clustering property is one of the differences between many real-world (i.e., complex) networks and random graphs. In many real-world networks we see that if there is a connection between node A and node B and between B and node C, then there is a high probability that A and C are also connected. In the language of social networks one would say: the friend of your friend is likely to also be your friend. So we have triangles in the network. One way to mathematically quantify this property is to define the *clustering coefficient* c_i of node i as the fraction of pairs of neighbors of node i that are also neighbors of each other. Suppose that node i has k_i edges that connect this node to k_i other nodes. Clearly, at most $k_i(k_i-1)/2$ edges can exist among them, and this occurs in a complete graph when every neighbor of node i is connected to every other neighbor of node i. The clustering coefficient c_i of node i is then defined as the ratio of the number e_i of edges that actually exist among these k_i nodes to the total possible number $k_i(k_i-1)/2$ of edges; that is,

$$c_i = \frac{e_i}{k_i(k_i-1)/2} = \frac{2e_i}{k_i(k_i-1)} \tag{10.18}$$

The clustering coefficient C of the whole network is the average of c_i over all i. Clearly, $C \le 1$ and $C=1$ if and only if every node in the network connects to every other node; that is, the network can be modeled by a complete graph. In a completely random network consisting of N nodes, $C \sim 1/N$, which is very small compared to those for most real networks. Thus, most real complex networks are not completely random. In fact, in most complex networks nodes with larger degrees tend to be connected; they are said to be *assortative*. For example, people who have a lot of contacts tend to know other people with many contacts.

10.9.3   Degree Distribution

As defined earlier, the degree k_i of a node i is the total number of its connections. Thus, the larger the degree of a node, the "more important" the node is in a network. The spread of node degrees over a network is characterized by a distribution function $P[k]$, which is the probability that a randomly selected node has exactly k edges. In the past few years, many empirical results have shown that for most large-scale real networks the degree distribution can be described by a power law of the form

$$P[k] \sim k^{-\gamma} \qquad \gamma > 0 \tag{10.19}$$

This power-law distribution falls off more gradually than an exponential distribution, allowing for a few nodes of very large degree to exist. Because these power-laws are free of any characteristic scale, such a network with a power-law degree distribution is called a *scale-free network*.

10.9.4   Percolation and Network Resilience

One of the major concerns in complex networks is *resilience*. Network resilience refers to the effect on the network of removal of its nodes and/or edges. Most networks rely on the connections between the nodes to function, and with the removal of nodes the path lengths in the network will increase. At some point some of the paths will no longer exist.

Percolation can be used to study the vulnerability or robustness of the connectivity under random or intentional breakdown of fractions of the network. Specifically, the problem of network resilience to random node failures can be described by a percolation process on the graph modeling the network. It is used to answer such questions as how large the connected components are and the average distance between nodes in the components if a given fraction of nodes or edges is removed from a network.

Percolation and network resilience are related by observing that the random breakdowns (failures) of network nodes more or less correspond to the classical site percolation problem, that is, a vertex of the network is present with probability $p = 1 - q$. Thus, just as percolation is studied by adding vertices (in site percolation) or edges (in bond percolation) until a giant component emerges, network resilience is studied by incrementally removing network vertices (in site percolation) or edges (in bond percolation). As these components are removed from the network, the average path length will increase until ultimately the giant component disintegrates. How fast the giant component disintegrates with the removal of network components depends on the pattern of component removal and the type of network.

The resilience to random failure (or removal) of nodes in a network is equivalent to a site percolation problem. Attacks (failures) in a network can be random or intentional (also called targeted). In random failure of links, each edge is removed with probability $q = 1 - p$. A targeted attack causes severe damage to the network with the removal of the fewest edges. Usually, it attempts to remove edges that are most likely to break apart the network or lengthen the average shortest path; one example is targeting edges with high *betweenness*, where betweenness is a measure of a node's importance to the network and is equal to the number of shortest paths from all nodes to all others that pass through that node.

Consider a node with an initial degree distribution $P[k_0]$, which is the probability that a node has k_0 incident edges. Let $P[k]$ be the probability that the node has $k < k_0$ incident edges after the random removal of nodes. Since $\binom{k_0}{k}$ is the number of ways to pick the surviving k nodes among the k_0 initial neighbors of the node, we have that

$$P[k] = \sum_{k_0 = k}^{\infty} P[k_0] \binom{k_0}{k} p^k (1-p)^{k_0 - k} \tag{10.20}$$

Another fact about network resilience is that scale-free networks are resilient with respect to random attacks. However, they are not resilient against targeted attacks. The removal of a successfully chosen small number of nodes may cause many more nodes (their neighbors) to be disconnected from the network. Also, in power grid structural networks, each node has a load and a capacity that specify how much load it can tolerate. When a node is removed from the network its load is distributed to the remaining nodes. If the load of a node exceeds its capacity, then the node fails resulting in a *cascading failure* or *avalanche* in which the redistribution of load pushes another node or edge over its threshold and causes it to fail, leading to further redistribution.

Further information on percolation in complex networks can be obtained from Cohen and Havlin (2010) and Dorogovtsev (2010).

10.10 SUMMARY

Percolation theory studies the deterministic propagation of a fluid on a random medium. It offers a good theoretical framework to study the behavior of complex systems and their characteristic phase transition phenomenon. The theory was introduced by Hammersley and Broadbent to model the deterministic propagation of a fluid through a random medium. It has been successfully applied to model complex systems in statistical physics, economics and social networks. It is also used in hydrology and fractal mathematics.

Percolation theory represents one of the simplest models of a disordered system. We say that there is percolation in the system when there is appearance of the giant (or infinite) cluster. Associated with percolation is the phase transition phenomenon, which is the abrupt change in the state of a system around a critical value of a key parameter. While the theory is closely related to random graph theory, percolation studies are usually carried out via simulation because of the complexity of the process.

In this chapter, we have discussed the essential principles of percolation and different models of percolation. These models include continuum percolation, BP, DP, first-passage percolation, explosive percolation, and percolation in complex networks. More information on the subject can be obtained from the cited references.

REFERENCES

Achlioptas, D., R.M. D'Souza and J. Spencer (2009). "Explosive Percolation in Random Networks," *Science*, vol. 323, pp.1453–1455.

Addison, P.S. (1997). *Fractals and Chaos: An Introductory Course*, Institute of Physics Publishing, Bristol, UK.

Adler, J. and A. Aharony (1988). "Diffusion Percolation I: Infinite Time Limit and Bootstrap Percolation," *Journal of Physics A: Mathematical and General*, vol. 21, pp. 1387–1404.

Argyrakis, P. and K.W. Kehr (1992). "Mean Number of Distinct Sites Visited by Correlated Walks II: Disordered Lattices," *Journal of Chemical Physics*, vol. 97, pp. 2718–2723.

Bandyopadhyay, S., E.J. Coyle and T. Falck (2006). "Stochastic Properties of Mobility Models in Mobile Ad Hoc Networks," *Proceedings of the 40th Annual Conference on Information Sciences and Systems*, pp. 1205–1211.

Barber, M.N. and B.W. Ninham (1970). *Random and Restricted Walks: Theory and Applications*, Gordon and Breach Science Publishers, New York.

Benassi, A., S. Jaffard and D. Roux (1997). "Elliptic Gaussian Random Processes," *Revista Matematica Iberoamericana*, vol.13, pp. 19–90.

Benhamou, S. (2007). "How Many Animals Do the Levy Walk?" *Ecology*, vol. 88, pp. 1962–1969.

Benth, F.E. (2004). *Option Theory with Stochastic Analysis: An Introduction to Mathematical Finance*, Springer, New York.

Bohm, W. (2000). "The Correlated Random Walk with Boundaries: A Combinatorial Solution," *Journal of Applied Probability*, vol. 37, pp. 470–479.

Elements of Random Walk and Diffusion Processes, First Edition. Oliver C. Ibe.
© 2013 John Wiley & Sons, Inc. Published 2013 by John Wiley & Sons, Inc.

Bohman, T. and A. Frieze (2001). "Avoiding a Giant Component," *Random Structures Algorithms*, vol. 19, pp. 75–85.

Bollobas, B. (1984). "The Evolution of Random Graphs," *Transactions of the American Mathematical Society*, vol. 286, pp. 257–274.

Bouchaud, J.P. and A. Georges (1990). "Anomalous Diffusion in Disordered Media: Statistical Mechanics, Models and Physical Applications," *Physics Reports*, vol. 195, pp. 127–293.

Broadbent, S.R. and J.M. Hammersley (1957). "Percolation Processes, I: Crystals and Mazes," *Proceedings of the Cambridge Philosophical Society*, vol. 53, pp. 629–641.

Brockmann, D. and L. Hufnagel (2007). "Front Propagation in Reaction-Superdiffusion Dynamics: Taming Lévy Flights with Fluctuations," *Physical Review Letters*, vol. 98, p. 178301.

Brockmann, D., L. Hufnagel and T. Geisel (2006). "The Scaling Laws of Human Travel," *Nature*, vol. 439, pp. 462–465.

Buchanan, M. (2008). "Ecological Modeling: The Mathematical Mirror to Animal Nature," *Nature*, vol. 453, pp. 714–716.

Burnecki, K., J. Janczura, M. Magdziarz and A. Weron (2008). "Can One See a Competition Between Subdiffusion and Levy Flights: A Case of Geometric-Stable Noise," *Acta Physica Polonica B*, vol. 39, pp. 1043–1054.

Byers, J.A. (2001). "Correlated Random Walk Equations of Animal Dispersal Resolved by Simulation," *Ecology*, vol. 82, pp. 1680–1690.

Capasso, V. and D. Bakstein (2005). *An Introduction to Continuous-Time Stochastic Processes: Theory, Models and Applications to Finance, Biology and Medicine*, Birkhauser, Boston, Massachusetts.

Chalupa, J., P.L. Leath and G.R. Reich (1979). "Bootstrap Percolation on a Bethe Lattice," *Journal of Physics C: Solid State Physics*, vol. 12, pp. L31–L35.

Chechkin, A.V., R. Metzler, V.Y Gonchar, J. Klafter and L.V. Tanatarov. (2003). "First Passage and Arrival Densities for Levy Flights and the Failure of the Method of Images," *Journal of Physics A: Mathematical and General*, vol. 36, pp. L537–L544.

Codling, E.A. (2003). "Biased Random Walks in Biology," PhD Thesis, University of Leeds, UK.

Codling, E.A., M.J. Plank, and S. Benhamou (2008). "Random Walk Models in Biology," *Journal of Royal Society Interface*, vol. 5, pp.813–834.

Cohen, R. and S. Havlin (2010). *Complex Networks: Structure, Robustness and Function*, Cambridge University Press, Cambridge, UK.

Coppersmith, D. and P. Diaconis (1987). "Random Walk with Reinforcement," Unpublished manuscript.

Dasgupta, A. (1998). "Fractional Brownian Motion: Its Properties and Applications to Stochastic Integration," Ph.D. Thesis, University of North Carolina.

Davis, B. (1990). "Reinforced Random Walk," *Probability Theory and Related Fields*, vol. 84, pp. 203–229.

de Gennes, P.G. (1976). "La Percolation: Un Concept Unificateur," *La Recherche*, vol. 7, pp. 919–927.

Diethelm, K. (2010). *The Analysis of Fractional Differential Equations*, Springer, New York.

Dorogovtsev, S.N. (2010). *Lectures on Complex Networks*, Oxford University Press, Oxford, UK.

Edwards, A.M., R.A. Phillips, N.W. Watkins, M.P. Freeman, E.J. Murphy, V. Afanasyev, S.V. Buldyrev, M.G.E. da Luz, E.P. Raposo, H.E. Stanley and G.M. Viswanathan (2007). "Revisiting Levy Flight Search Patterns of Wandering Albatrosses, Bumblebees and Deer," *Nature*, vol. 449, pp.1044–1048.

Escande, D.F. and F. Sattin (2007). "When Can the Fokker-Planck Equation Describe Anomalous or Chaotic Transport?" *Physical Review Letters*, vol. 99, p. 185005.

Gilbert, E.N. (1961). "Random Plane Networks," *Journal of the Society for Industrial and Applied Mathematics*, vol. 9, pp. 533–543.

Gillespie, D.T. (1996). "The Mathematics of Brownian Motion and Johnson Noise," *American Journal of Physics*, vol. 64, pp.225–240.

Gillis, J. (1955). "Correlated Random Walk," *Proceedings of the Cambridge Philosophical Society*, vol. 51, pp. 639–651.

Goldstein, S. (1951). "On Diffusion by Discontinuous Movements, and on the Telegraph Equation," *Quarterly Journal of Mechanics*, vol. 4, pp. 129–156.

Gonzales, M.C., C.A. Hidalgo and A.-L. Barabasi (2008). "Understanding Individual Human Mobility Patterns," *Nature*, vol. 453, pp. 779–782.

Hanneken, J.W. and D.R. Franceschetti (1998). "Exact Distribution Function for Discrete Time Correlated Random Walks in One Dimension," *Journal of Chemical Physics*, vol. 109, pp. 6533–6539.

Haus, J.W and K.W. Kehr (1987). "Diffusion in Regular and Disordered Lattices," *Physics Reports*, vol. 150, pp. 263–416.

Havlin, S. and D. Ben-Avraham (2002). "Diffusion in Disordered Media," *Advances in Physics*, vol. 51, pp. 187–292.

Helmstetter, A. and D. Sornette (2002). "Diffusion of Earthquake Aftershocks, Omori's Law, and Generalized Continuous-time Random Walk Models," *Physics Review E*, vol. 66, p. 061104.

Hemmer, S. and P. C. Hemmer (1984). "An Average Self-avoiding Random Walk on the Square Lattice Lasts 71 Steps," *Journal of Chemical Physics*, vol. 81, pp. 584–585.

Herrmann, R. (2011). *Fractional Calculus: An Introduction for Physicists*, World Scientific Publishing Co., Singapore.

Huang, K. (1987). *Statistical Mechanics*, 2nd ed. John Wiley & Sons, New York.

Hughes, B.D. (1995). *Random Walk and Random Environments*, Vol. 1. Oxford University Press, Oxford, UK.

Ibe, O.C. (2005). *Fundamentals of Applied Probability and Random Processes*, Academic Press, New York.

Jain, G.C. (1971). "Some Results in a Correlated Random Walk," *Canadian Mathematical Bulletin*, vol. 14, pp. 341–347.

Jonsen, I.D., J.M. Flemming and R.A. Myers (2005). "Roburst State-Space Modeling of Animal Movement Data," *Ecology*, vol. 86, pp. 2874–2880.

Kareiva, P.M. and N. Shigesada (1983). "Analyzing Insect Movement as a Correlated Random Walk," *Oecologia*, vol. 56, pp. 234–238.

Kehr, K.W. and P. Argyrakis (1986). "Mean Number of Distinct Sites Visited by Correlated Walks I: Perfect Lattices," *Journal of Chemical Physics*, vol. 84, pp. 5816–5823.

Keller, J.B. (2004). "Diffusion at Finite Speed and Random Walk," *Proceedings of the National Academy of Sciences*, vol. 101, pp. 1120–1122.

Kilbas, A., H.M. Srivastava and J.J. Trujillo (2006). *Theory and Applications of Fractional Differential Equations*, Elsevier, Amsterdam, The Netherlands.

Klebaner, F.C. (2005). *Introduction to Stochastic Calculus with Applications*, 2nd ed. Imperial College Press, London.

Koren, T., J. Klafter and M. Magdziarz (2007a). "First Passage Times of Lévy Flights Coexisting with Subdiffusion," *Physical Review E*, vol. 76, p. 031129.

Koren, T., A.V. Chechkin and J. Klafter (2007b). "On the First Passage Time and Leapover Properties of Lévy Motions," *Physica A*, vol. 379, pp. 10–22.

Koren, T., M.A. Lombolt, A.V. Chechkin, J. Klafter, and R. Metzler (2007c). "Leapover Lengths and First Passage Time Statistics for Levy Flights," *Physical Review Letters*, vol. 99, p. 160602.

Lal, R. and U.N. Bhat (1989). "Some Explicit Results for Correlated Random Walks," *Journal of Applied Probability*, vol. 27, pp. 756–766.

Mandelbrot, B.B. (1982). *The Fractal Geometry of Nature*, W.H. Freeman and Co., San Francisco

Mandelbrot, B.B. and J.W. van Ness (1968). "Fractional Brownian Motions, Fractional Noises and Applications," *SIAM Review*, vol. 10, pp. 422–437.

Mantegna, R.N. and H.E. Stanley (1994). "Stochastic Processes with Ultraslow Convergence to a Gaussian: the Truncated Levy Flight," *Physical Review Letters*, vol. 73, pp. 2946–2949.

Mantegna, R.N. and H.E. Stanley (2000). *An Introduction to Econophysics*, Cambridge University Press, Cambridge, UK.

Masoliver, J. and M. Montero (2003). "Continuous-time Random-walk Model for Financial Distributions," *Physical Review E*, vol. 67, p. 021112.

Masoliver, J., M. Montero, J. Perello and G.H. Weiss (2006). "The Continuous Time Random Walk Formalism in Financial Markets," *Journal of Economic Behavior & Organization*, vol. 61, pp. 577–598.

Metzler, R. and J. Klafter (2000). "The Random Walk's Guide to Anomalous Diffusion: A Fractional Dynamics Approach," *Physics Reports*, vol. 339, pp. 1–77.

Metzler, R. and J. Klafter (2004). "The Restaurant at the end of the Random Walk: Recent Developments in the Description of Anomalous Transport by Fractional Dynamics," *Journal of Physics A: Mathematical and General*, vol. 37, pp. R161–R208.

Mikosch, T., S. Resnick, H. Rootzen and A. Stegeman (2002). "Is Network Traffic Approximated by Stable Levy Motion or Fractional Brownian Motion?" *Annals of Applied Probability*, vol. 12, pp. 23–68.

Mohan, C. (1955). "The Gambler's Ruin Problem with Correlation," *Biometrika*, vol. 42, pp. 486–493.

Montroll, E.W. and M.F. Shlensinger (1984). In *Nonequilibrium Phenomena II: From Stochastics to Hydrodynamics*, J.L. Lebowitz and E.W. Montroll (editors), North-Holland, Amsterdam, The Netherlands, pp. 1–121.

Montroll, E.W. and G.H. Weiss (1965). "Random Walks on Lattice II," *Journal of Mathematical Physics*, vol. 6, pp. 167–181.

Narayanan, R.G.L. (2011). "An Architecture for Disaster Recovery and Search and Rescue Wireless Networks," Ph.D. Thesis, Department of Electrical and Computer Engineering, University of Massachusetts Lowell.

Norros, I. (1995). "On the Use of Fractional Brownian Motion in the Theory of Connectionless Networks," *IEEE Journal on Selected Areas in Communications*, vol. 13, pp. 953–962.

Oksendale, B. (2005). *Stochastic Differential Equations*, 6th ed. Springer, New York.

Oldham, K.B. and J. Spanier (1974). *The Fractional Calculus: Theory and Applications of Differentiation and Integration to Arbitrary Order*, Academic Press, New York. (Dover Publication edition, 2006).

Othmer, H.G., S.R., Dunbar, and W. Alt (1988). "Models of Dispersal in Biological Systems," *Journal of Mathematical Biology*, vol. 26, pp. 263–298.

Paoletti, M.S., C.R. Nugent, T.H. Solomon (2006). "Synchronization of Oscillating Reactions in an Extended Fluid System," *Physical Review Letters*, vol. 96, p. 124101.

Peltier, R.F. and J.L. Vehel (1995). "Multifractional Brownian Motion: Definition and Preliminary Results," *INRIA Technical Report No.* 2645.

Pemantle, R. (2007). "A Survey of Random Processes with Reinforcement," *Probability Surveys*, vol. 4, pp. 1–79.

Podlubny, I. (1999). *Fractional Differential Equations*, Academic Press, San Diego, California.

Porto, M. and H.E. Roman (2002). "Autoregressive Processes with Exponentially Decaying Probability Distribution Functions: Application to Daily Variations of a Stock Market Index," *Physical Review E*, vol. 65, p. 046149.

Renshaw, E. and R. Henderson (1981). "The Correlated Random Walk," *Journal of Applied Probability*, vol. 18, pp. 403–414.

Rhee, I., M. Shin, S. Hong, K. Lee and S. Chong (2008). "On the Levy-walk Nature of Human Mobility," *Proceedings of the IEEE INFOCOM*, pp. 1597–1605.

Rogers, L.C.G. (1997). "Arbitrage with Fractional Brownian Motion," *Mathematical Finance*, vol. 7, pp. 95–105.

Roman, H.E. and M. Porto (2008). "Fractional Derivatives of Random Walks: Time Series with Long-time Memory," *Physical Review E*, vol. 78, p. 031127.

Scalas, E. (2006a). "Five Years of Continuous-time Random Walks in Econophysics," in *The Compplex Networks of Economic Interactions: Essays in Agent-Based Economics and Econophysics*, A. Namatame, T. Kaizouji and Y. Aruga (editors), Springer, Tokyo, pp. 1–16.

Scalas, E. (2006b). "The Application of Continuous-time Random Walks in Finance and Economics," *Physica A*, vol. 362, pp. 225–239.

Scher, H. and E.W. Montroll (1975). "Anomalous Transition-Time Dispersion in Amorphous Solids," *Physical Review B*, vol. 12, pp. 2455–2477.

Seth, A. (1963). "The Correlated Unrestricted Random Walk," *Journal of the Royal Statistical Society, Series B*, vol. 25, pp. 394–400.

Slade, G. (1994). "Self-Avoiding Walks," *The Mathematical Intelligencer*, vol. 16, pp. 29–35.

Song, C., S. Havlin, and H. A. Makse (2006), "Origins of Fractality in the Growth of Complex Networks," *Nature Physics*, vol. 2, pp. 275–281.

Sottinen, T. (2001). "Fractional Brownian Motion, Random Walks and Binary Market Models," *Finance and Stochastics*, vol. 5, pp. 343–355.

Sparre Andersen, E. (1953). "On Fluctuations of Sums of Random Variables," *Mathematica Scandinavica*, vol. 1, pp. 263–285.

Sparre Andersen, E. (1954). "On Fluctuations of Sums of Random Variables II," *Mathematica Scandinavica*, vol. 2, pp. 194–223.

Stanley, H.E., L.A.N. Amaral, D. Canning, P. Gopikrishnan, Y. Lee and Y. Liu (1999). "Econophysics: Can Physicists Contribute to the Science of Economics?" *Physica A*, vol. 269, pp. 156–169.

Steele, M.J. (2001). *Stochastic Calculus and Financial Applications*, Springer, New York.

Tambasco, M., M. Eliasziw and A.M. Magliocco (2010). "Morphologic Complexity of Epithelial Architecture for Predicting Invasive Breast Cancer Survival," *Journal of Translational Medicine*, vol. 8, p. 140.

Temperley, H.N.V. (1956). "Combinatorial Problems Suggested by the Statistical Mechanics of Domains and Rubber-like Molecules," *Physical Review*, vol. 103, pp. 1–16.

Vainstein, M.H., L.C. Lapas and F.A. Oliveira (2008), "Anomalous Diffusion," *Acta Physica Polonica B*, vol. 39, pp. 1273–1281.

Van Duen, J. and R. Cools (2006). "Algorithm 858: Computing Infinite Range Integrals of an Arbitrary Product of Bessel Functions," *ACM Transactions on Mathematical Software*, vol. 32, pp.580–596.

Van Duen, J. and R. Cools (2008). "Integrating Products of Bessel Functions with an Additional Exponential or Rational Factor," *Computer Physics Communications*, vol. 178, pp. 578–590.

Verdier, P.H. and E.A. DiMarzio (1969). "On the Mean Dimensions of Restricted Random Walks," *Journal of Research of the National Bureau of Standards – B: Mathematical Sciences*, vol. 73B, pp. 45–46.

Viswanathan, G.M., S.V. Buldyrev, S. Havlin, M.G.E. da Luz, E.P. Raposo and H.E. Stanley (1999). "Optimizing the Success of Random Searches," *Nature*, vol. 401, pp. 911–914.

Weiss, G.H., J.M. Porra and J. Masoliver (1998). "Statistics of the Depth Probed by CW Measurements of Photons in a Turbid Medium," *Physical Review E*, vol. 58, pp. 6431–6439.

Wilf, H.S. (1990). *Generatingfunctionology*, Academic Press, Boston, Massachusetts, p. 50.

Zaslavsky, G.M. (2002). "Chaos, Fractional Kinetics, and Anomalous Transport," *Physics Reports*, vol. 371, 461–580.

INDEX

absorbing barriers, 50, 52, 86, 101
absorbing state, 34
Achlioptas process, 241
adapted process, 145
alternating random walk (ARW), 104
   Bernoulli trials, 115
   multinomial distribution, 115–16
   probability of eventual return, 117
   Stirling's approximation, 117
anomalous diffusion
   asymptotic behavior, 218
   CTRW, 216
   master equation states, 217
   normal diffusion equation, 215
   scaling index, 216
   space-fractional case, 220–221
   time-fractional case, 219–20
arc-sine law, 71
arithmetic Brownian motion, 188
autocorrelation function, 22–4,
     182, 212

backward diffusion equation, 162,
     164, 167
ballistic diffusion, 216
Ballot problem, 65–7
Bernoulli distribution, 11–12
Bernoulli random graphs, 226–8
Bernoulli random walk, 46
binomial distribution, 12
birth and death processes, 36–8
Black–Scholes model, 159
Bohman–Frieze rule, 242
bond percolation, 226, 228–9, 245
Boolean model, 235–6
bootstrap percolation (BP), 237
box-count fractal dimension, 184
Brownian bridge, 136–7
Brownian motion *see also* stochastic
     calculus
   boundary conditions, 165
   classical, 132
   continuous-time stochastic processes, 131

Elements of Random Walk and Diffusion Processes, First Edition. Oliver C. Ibe.
© 2013 John Wiley & Sons, Inc. Published 2013 by John Wiley & Sons, Inc.

Brownian motion *see also* stochastic
   calculus (*cont'd*)
  diffusion equation, 164–5
  with drift, 132, 167, 188
  fractional calculus
   condition, 211
   Hurst index, 211
   model financial markets, 213
   self-similarity, 212
   stochastic integral, 210–211
   types of processes, 212
  first passage time, 133–4
  functions, 166
  geometric, 137
  Langevin equation, 137–40
  as Markov process, 132–3
  as martingale, 133
  maximum value, 135
  method of characteristic equation, 165–6
  pollen particles suspended in fluid, 129
  properties, 131
  sample function, 132
  standard, 132
  tied-down, 136–7
  time in interval, 135–6

Cantor set, 185
Caputo fractional derivative, 205–6
central limit theorems, 18–19
  generalized, 175–7
  stable distribution, 180–182
central moments, 4
Chapman–Kolmogorov equations, 32, 36
characteristic exponent, 178
Chebyshev inequality, 17
classical Brownian motion, 132
cliques, 226, 227
complex networks
  average path length, 243
  clustering coefficient, 243–4
  degree distribution, 244
  resilience, 244–5
  scale-free nature, 243
  small-world phenomenon, 242
compound Poisson process
  descriptions, 28–9
  infinite divisibility, 187–8
  Levy processes, 187–8
conditional probability method, 66–7

connected graphs, 74–5, 77, 79–80
continuous random variables, 3–4
continuous-time Gauss–Markov
   process, 151
continuous-time Markov chains, 35–6
continuous-time random walk (CTRW)
  characteristic function, 91–2
  vs. discrete-time, 90
  Levy walk, 192–5
  master equation, 92–4
  random variables, 90
  uncoupled and coupled CTRW, 91
  waiting times, 90
continuum percolation models
  Boolean model, 235–6
  point process, 235
  random connection model, 236
convection-diffusion equation, 162
convolution integral, 10
correlated random walks (CRW)
  bivariate Markov chain, 85
  characteristic functions, 87
  definition, 85
  diffusion processes, 167–70
  eigenvalues, 88
  eigenvectors, 89
  model animal and insect movement, 90
  quasi-birth-and-death process, 86
  state-transition diagrams, 86
  two-state Markov chain, 87
  win/loss, 85
correlation coefficient, 9
correlation function, 231, 234
counting process, 25–6, 61, 222
covariance, 9, 136, 152, 211, 221
cumulative distribution functions
   (CDF), 2
  Bernoulli distribution, 11
  binomial distribution, 12
  Brownian motion, 134
  Cauchy distribution, 179
  central limit theorem, 18–19
  continuous random variables, 10
  discrete random variable, 3
  exponential distribution, 13
  first-passage percolation, 239
  Gaussian distribution, 179
  geometric distribution, 12
  infinite divisibility property, 186

Levy distribution, 179
Markov process, 29
normal distribution, 14, 16
Poisson distribution, 13
Poisson process, 27
strict-sense stationary process, 23

delay time, 22
diffusion percolation (DP)
    vs. bootstrap, 239
    cluster growth, 237
    mean square displacement, 238
    mobility, 238–9
diffusion processes
    approximation of Pearson random walk, 173
    Black–Scholes model, 159
    Brownian motion, 164–7
    concentration gradient, 158
    correlated random walk and telegraph
        equation, 167–70
    definition, 159
    on 1D random walk, 160–164
    at finite speed, 170–171
    Markov process, 159–60
    symmetric 2D lattice random walk, 171–2
diffusivity, 161
digraph, 75–6
directed graph, 75–6
disconnected graphs, 74
discrete random variables, 3, 6–8, 13, 28
discrete-time Markov chains, 30–31
discrete-time random process, 22
doubly stochastic matrix, 35, 123,
        125, 126, 128
doubly stochastic Poisson process, 28
dwell time, 36

edge-reinforced random walk (ERRW), 97
    Polya's urn model, 94–7
    weighted graph, 94
effective resistance
    computation of, 84–5
    escape probability, 83
    nodes, 82
    rules, 83–4
electric networks
    conductance, 81
    effective resistance, 82–5
    harmonic functions, 82

nodes, 81–2
resistance, 80
enhanced diffusion, 216
Erdos-Renyi (ER) random graph, 240–241
ergodic chain, 34
ergodic states, 34
escape probability, 83
explosive percolation
    Bohman–Frieze rule, 242
    component structure, 240
    Erdos-Renyi model, 240–241
    giant component, 240–241
    product rule, 241
    sum rule, 242
exponential distribution, 13–14
exponential random walk *see* geometric
        random walk

fast mobility model *see* Levy flights
Fick's law, 162–3
filtered probability space, 145
first-passage percolation, 239–40
first passage time (FPT)
    Brownian motion, 133–4
    Levy flights, 190
    symmetric random walk, 59–65
Fokker–Planck equation, 161
forward diffusion equation, 161
Fourier–Laplace transform, 92–3, 193–4
Fourier transform, 5, 24, 193–5
fractals
    Cantor set, 185
    definition, 183
    dimension, 184–5
    Levy flight, 190
    Sierpinski gasket, 183–4
fractional calculus
    Brownian motion, 210–213
    definition, 196
    derivatives, 201–2
    differential equations, 207–10
    diffusion (*see* anomalous diffusion)
    gamma function, 197–8
    Gaussian noise, 221–2
    integrals, 203
    integro-differentials, 203–6
    Laplace transform, 200–201
    MBM, 213
    memory-dependent phenomena, 196

fractional calculus (*cont'd*)
    Mittag–Leffler functions, 198–200
    Poisson process, 222–4
    random walk, 213–15

gambler's ruin
    duration of game, 55–6
    Levy flights, 190
    random walk with stay, 56
    ruin probability, 52–4
gamma-driven OU process, 152
gamma function, 181, 197–8, 202, 204
Gaussian noise, fractional calculus, 221–2
Gaussian processes, 38, 151, 154, 211, 221
Gaussian random walk, 99
generalized central limit theorem
    black hole of statistics, 176
    power-law, 176–7
    requirements, 176
    sums of random variables, 175
geometric Brownian motion, 137
geometric distribution, 12–13
geometric random walk, 98–9
graphs
    complete graph, 226, 227
    connected, 74–5, 77, 79–80
    definition, 226
    directed, 75–6
    disconnected, 74
    with loop, 74–5
    multigraph, 74–5
    random graph, 226–8
    simple, 74–5
    undirected, 76–9
Grunwald–Letnikov (GL) fractional
    derivative, 206

holding time, 36
homogeneous Poisson process, 27

infinite divisibility, 186–7
    Brownian motion with drift, 188
    compound Poisson process, 187–8
    Poisson process, 187
irreducible Markov chain, 33
Ito integral, 145–6
Ito's formula, 147–8, 152

k-core percolation, 237

lag time, 22
Langevin equation
    assumption, 137
    Brownian particle, 140
    equipartition theorem, 139
    frictional force, 138
    mean-square displacement, 137
Laplace transform, 92, 193–5, 200–201,
    206–7, 209, 217–20, 223
Levy distribution, 185–6
Levy flights, 190 *see also* fractals
    application, 189
    definition, 188
    first passage time, 190
    fractals, 190
    leapover properties, 190–191
    model phenomena, 188
    superdiffusive, 189
    TLF, 191
Levy processes *see also* stochastic process
    Brownian motion with drift, 188
    compound Poisson process, 187–8
    conditions, 186
    infinite divisibility property, 186–7
    Poisson process, 187
Levy walk
    coupled CTRW, 192–5
    vs. Levy flight, 191–2
    truncated Levy walk, 195
limiting-state probabilities, 34–5
limit theorems
    central limit, 18–19
    Chebyshev inequality, 17
    laws of large numbers, 17–18
    Markov inequality, 16

Markov chains
    continuous-time, 35–6
    discrete-time, 30–31
    noncontinuing random walk, 126
    nonreversing and noncontinuing
        properties, 127
    nonreversing random walk, 122–3
    1D random walk, 49
    time homogeneous, 35
Markov inequality, 16
Markov processes
    birth and death processes, 36–8
    classification of states, 33–4

continuous-time Markov chain, 35–6
discrete-time Markov chains, 30–31
doubly stochastic matrix, 35
first-order process, 29
k-step state transition probability, 31–2
limiting-state probabilities, 34–5
second-order process, 29
sequence of games, 29
state-transition diagrams, 32
state transition probability matrix, 31
types, 29–30
Markov property, 29
Martingales, 38–40
  Brownian motion, 133
  stopping times, 40
  symmetric random walk, 49–50
master equation, 92–4
mean-reverting Ornstein–Uhlenbeck
      process, 155–7
mean-square displacement (MSD)
  diffusion percolation, 238
  Langevin equation, 137
  nonreversing random walk, 126
  1D random walk, 50–52
  Pearson random walk, 105–7
  symmetric 2D random walk, 114
Mittag–Leffler functions, 198–200
moment-generating property
  characteristic function, 6
  s-transform, 7
  z-transform, 8–9
multifractional Brownian motion
      (MBM), 213
multigraph, 74–5

network resilience, 244–5
nonanticipating process, 145
noncontinuing random walk
      (NCRW), 126–7
nonhomogeneous Poisson process, 27
nonrecurrent state, 33
nonreversing and noncontinuing random
      walk (NRNCRW), 127–8
nonreversing random walk (NRRW), 99
  amount of memory, 121
  Markov chain, 122–3
  mean-square displacement, 126
  NCRW, 126–7
  NRNCRW, 127–8

turn angle, 122
z-transform, 123–5
normal distribution, 14–16
null recurrent state, 34

one-dimensional (1D) random walk
  alternative derivation, 163–4
  arc-sine law, 71
  backward diffusion equation, 162
  ballot problem and reflection
      principle, 65–7
  barriers, 50, 51
  Bernoulli random walk, 46
  convection-diffusion equation, 162
  CRW, 85–90
  CTRW, 90–94
  definition, 45
  density fluctuations, 160
  and electric networks
    conductance, 81
    effective resistance, 82–5
    harmonic functions, 82
    nodes, 81–2
    resistance, 80
  ERRW, 94–7
  exchangeability property, 73
  Fick's law, 162–3
  first passage times
    generating function, 59–61
    hitting time, 64–5
    reflection principle, 61–4
  first return to origin, 57–9
  forward diffusion equation, 161
  gambler's ruin
    duration of game, 55–6
    ruin probability, 52–4
  on graph
    connection notion, 74
    degree of vertex, 74–5
    directed graph, 75–6
    proximity measures, 75
    subgraph, 74
    undirected graph, 76–9
    vertices, 73–4
    weighted graph, 80
  with hesitation, 56–7
  Markov chain, 49
  maximum of, 72–3
  mean-square displacement, 50–52

one-dimensional (1D) random walk (*cont'd*)
  models
    Gaussian random walk, 99
    geometric random walk, 98–9
    with memory, 99–100
  occupancy probability, 46–8
  returns to origin, 67–71
  sample path, 45
  sequence of Bernoulli trials, 44
  skip-free property, 46
  symmetric random walk, 46, 49–50
Ornstein–Uhlenbeck (OU) process
  continuous-time Gauss–Markov
      process, 151
  factors, 151
  first alternative solution method, 153
  independent increments, 152
  mean-reverting process, 155–7
  SDE solution, 152–3
  second alternative solution method, 154–5
oscillation differential equation, 208–9

pair connectivity, 231
Pearson random walk
  diffusion approximation, 173
  mean-square displacement, 105–7
  probability distribution, 107–10
percolation theory
  bonds, 226
  in complex networks, 242–5
  continuum percolation, 235–6
  diffusion percolation, 237–9
  explosive, 240–242
  first-passage, 239–240
  graph theory
    complete graph, 226, 227
    definition, 226
    random graph, 226–8
  k-core, 237
  on lattice, 228–9
    cluster configurations, 231–2
    cluster formation, 229–30
    correlation function, 231
    critical exponents, 234
    occupied and unoccupied sites, 232–3
    percolation probability, 233–4
    probability, 230–231
    size of clusters, 234–5
    threshold, 229

sites, 226
topology, 225
periodic state, 34
persistent random walks *see* correlated
    random walks (CRW)
persistent state, 33
Poisson distribution, 13
Poisson-driven OU process, 151
Poisson point process, 28
Poisson process
  compound Poisson process, 28–9
  definition, 26
  exponential distribution, 27
  fractional calculus, 222–4
  homogeneous and nonhomogeneous, 27
  infinite divisibility, 187
Poisson random graph, 228
Polya's urn model
  edge-reinforced random walk, 96–7
  exchangeability property, 96
  probability, 95–6
  self-reinforcing property, 95
  urn with red and black balls, 94–5
positive recurrent state, 34
power spectral density, 24
probability theory
  Bernoulli distribution, 11–12
  binomial distribution, 12
  covariance and correlation coefficient, 9
  exponential distribution, 13–14
  geometric distribution, 12–13
  limit theorems
    central limit theorem, 18–19
    Chebyshev inequality, 17
    laws of large numbers, 17–18
    Markov inequality, 16
  normal distribution, 14–16
  Poisson distribution, 13
  random variables, 1–5
  sums of independent random variables, 10
  transform methods, 5–9

random connection model, 236
random process *see* stochastic process
random sequence, 22
random variables
  continuous, 3–4
  CTRW, 90
  discrete, 3, 6–8, 13, 28

distribution functions, 2–3
domain, 2
expectations, 4
generalized central limit theorem, 175
moments, 4
probability theory, 1–5
range, 2
value, 1
variance, 4–5
recurrent random walk, 49
recurrent state, 33–4
reflecting barriers, 50, 86
reflection principle, 61–4
relaxation and oscillation fractional
    differential equations, 209–10
relaxation differential equation, 208
Riemann integral, 144
Riemann–Liouville (RL) fractional
    derivative, 204–5
ruin probability
  boundary condition, 53
  derivation, 54
  difference equation, 52–3
  state-transition diagram, 51, 52

scale-free network, 244
scale parameter, 178
self-affine *see* fractals
self-avoiding polygon (SAP), 118
self-avoiding random walk (SAW)
  Hammersely–Morton theorem, 119
  inequality, 119
  linear polymer molecules, 117
  mean-square length, 120
  random walk with memory, 99
  self-trapping, 117–18
  traps at 7th and 8th step, 121
  values on 2D square lattice, 118–20
self-similarity, 182
self-trapping walk, 117–18
shift parameter, 178
Sierpinski gasket, 183–4
simple graph, 74–5
site percolation, 226, 228–9, 230, 234, 245
skewness parameter, 178
slow mobility model *see* Levy walk
small-world networks, 243
space-fractional diffusion equation, 216
stability index, 178

stable distribution
  Cauchy distribution, 179
  definition, 177–8
  Gaussian distribution, 179
  generalized central limit theorem, 180–182
  leptokurtic distributions, 180
  Levy distributions, 179
  parameters, 178
standard Brownian motion, 132
state-transition diagrams
  CRW, 86
  gambler's ruin probability, 52
  graph, 76–8
  Markov process, 32
  1D random walk, 49
  random walk with stay, 56
stationary process
  strict-sense, 23–4
  WSS process, 24
Stirling's approximation
  alternating random walk, 117
  symmetric 2D random walk, 112
stochastic calculus
  dynamics of system, 143
  geometric Brownian motion, 148–51
  investor behavior and asset pricing, 145
  Ito integral, 145–6
  Ito's formula, 147
  OU process (*see* Ornstein–Uhlenbeck
    (OU) process)
  Riemann integral, 144
  SDE, 147–8
  stochastic differential, 146–7
  stochastic process, 144
  white noise, 144
stochastic differential equations (SDE), 147–8
stochastic process
  autocorrelation function, 22–3
  classification of, 22
  counting process, 25
  Gaussian process, 38
  independent increment process, 25
  Markov process (*see* Markov processes)
  martingales, 38–40
  mean of, 22
  parameter set, 21
  Poisson process, 26–8
  power spectral density, 24
  stationary increment process, 25

stochastic process (*cont'd*)
  strict-sense process, 23–4
  WSS, 24
stochastic Taylor series, 147
stopping times, 40
s-transform, 6–7 *see also* Laplace
    transform
strict-sense stationary process, 23–4
strong laws of large numbers, 17–18
superdiffusion, 216
suppressed diffusion, 216
symmetric random walk (SRW), 46, 110–112
  first passage times
    generating function, 59–61
    hitting time, 64–5
    reflection principle, 61–4
  as martingale, 49–50
  mean-square displacement, 114
  probability of eventual return, 113–14
  Stirling's approximation, 112
  2D lattice, 171–2
  two independent walk, 114–15

tail exponent, 178
tail index, 178
telegraph equation, 169–70
tied-down Brownian motion, 136–7
time-fractional diffusion equation, 216
time homogeneous Markov chains, 35
transform methods
  characteristic function, 5–6
  Fourier transform, 5, 24, 193–5
  joint Fourier–Laplace transform, 92–3,
    193–4

s-transform, 6–7 (*see also* Laplace
    transform)
types, 5
z-transform, 7–9
transient state, 33
transition density function, 159
trapping states, 34
truncated Levy flight (TLF), 191
truncated Levy walk (TLW), 195
two-dimensional (2D) random walk
  ARW, 115–17
  models, 104
  NRRW, 121–8
  Pearson random walk, 105–10
  SAW, 117–21
  SRW, 110–115

undirected graph
  checkerboard, 79
  multigraph, 76–8
  search technique, 76
  simple graph, 78–9
  state-transition diagram, 76–7
unrestricted CRW, 86

weak law of large numbers, 17–18
weighted graph, 80
  ERRW, 94
  1D random walk, 80
wide-sense stationary process, 24
Wiener process *see* Brownian motion

z-transform, 7–9